Magnolia

Medicinal and Aromatic Plants—Industrial Profiles

Individual volumes in this series provide both industry and academia with in-depth coverage of one major medicinal or aromatic plant of industrial importance.

Edited by Dr Roland Hardman

Magnolia

The genus *Magnolia*

Edited by
Satyajit D. Sarker
The Robert Gordon University,
Aberdeen, Scotland, UK
and Yuji Maruyama
Gunma University School of Medicine,
Gunma, Japan

CRC Press
Taylor & Francis Group
Boca Raton London New York

CRC Press is an imprint of the
Taylor & Francis Group, an **informa** business

A TAYLOR & FRANCIS BOOK

First published 2002 by Taylor & Francis

Published 2019 by CRC Press
Taylor & Francis Group
6000 Broken Sound Parkway NW, Suite 300
Boca Raton, FL 33487-2742

© 2002 by Taylor & Francis Group, LLC
CRC Press is an imprint of Taylor & Francis Group, an Informa business

First issued in paperback 2019

No claim to original U.S. Government works

ISBN 13: 978-0-367-45482-1 (pbk)
ISBN 13: 978-0-415-28494-3 (hbk)

Visit the Taylor & Francis Web site at
http://www.taylorandfrancis.com

and the CRC Press Web site at
http://www.crcpress.com

Typeset in 11/12pt Garamond 3 by Graphicraft Ltd., Hong Kong

Every effort has been made to ensure that the advice and information in this book is true and accurate at the time of going to press. However, neither the publisher nor the authors can accept any legal responsibility or liability for any errors or omissions that may be made. In the case of drug administration, any medical procedure or the use of technical equipment mentioned within this book, you are strongly advised to consult the manufacturer's guidelines.

British Library Cataloguing in Publication Data
A catalogue record for this book is available from the British Library

Library of Congress Cataloging in Publication Data
A catalogue record has been requested

Contents

Preface to the series

There is increasing interest in industry, academia and the health sciences in medicinal and aromatic plants. In passing from plant production to the eventual product used by the public, many sciences are involved. This series brings together information which is currently scattered through an ever increasing number of journals. Each volume gives an in-depth look at one plant genus, about which an area specialist has assembled information ranging from the production of the plant to market trends and quality control.

Many industries are involved such as forestry, agriculture, chemical, food, flavour, beverage, pharmaceutical, cosmetic and fragrance. The plant raw materials are roots, rhizomes, bulbs, leaves, stems, barks, wood, flowers, fruits and seeds. These yield gums, resins, essential (volatile) oils, fixed oils, waxes, juices, extracts and spices for medicinal and aromatic purposes. All these commodities are traded worldwide. A dealer's market report for an item may say "Drought in the country of origin has forced up prices".

Natural products do not mean safe products and account of this has to be taken by the above industries, which are subject to regulation. For example, a number of plants which are approved for use in medicine must not be used in cosmetic products.

The assessment of safe to use starts with the harvested plant material, which has to comply with an official monograph. This may require absence of, or prescribed limits of, radioactive material, heavy metals, aflatoxin, pesticide residue, as well as the required level of active principle. This analytical control is costly and tends to exclude small batches of plant material. Large scale contracted mechanised cultivation with designated seed or plantlets is now preferable.

Today, plant selection is not only for the yield of active principle, but for the plant's ability to overcome disease, climatic stress and the hazards caused by mankind. Such methods as *in vitro* fertilization, meristem cultures and somatic embryogenesis are used. The transfer of sections of DNA is giving rise to controversy in the case of some end-uses of the plant material.

Some suppliers of plant raw material are now able to certify that they are supplying organically-farmed medicinal plants, herbs and spices. The Economic Union directive (CVO/EU No. 2092/91) details the specifications for the *obligatory* quality controls to be carried out at all stages of production and processing of organic products.

Fascinating plant folklore and ethnopharmacology leads to medicinal potential. Examples are the muscle relaxants based on the arrow poison, curare, from species of *Chondrodendron*, and the anti-malarials derived from species of *Cinchona* and *Artemisia*. The methods of detection of pharmacological activity have become increasingly reliable and specific, frequently involving enzymes in bioassays and avoiding the use of laboratory animals. By using bioassay linked fractionation of crude plant juices or extracts, compounds can be specifically targeted which, for example, inhibit blood platelet aggregation, or have anti-tumour, or anti-viral, or any other required activity. With the assistance of robotic devices, all the members of a genus may be readily screened. However, the plant material must be *fully* authenticated by a specialist.

The medicinal traditions of ancient civilisations such as those of China and India have a large armamentarium of plants in their pharmacopoeias which are used throughout South-East Asia. A similar situation exists in Africa and South America. Thus, a very high percentage of the World's population relies on medicinal and aromatic plants for their medicine. Western medicine is also responding. Already in Germany all medical practitioners have to pass an examination in phytotherapy before being allowed to practise. It is noticeable that throughout Europe and the USA, medical, pharmacy and health related schools are increasingly offering training in phytotherapy.

Multinational pharmaceutical companies have become less enamoured of the single compound magic bullet cure. The high costs of such ventures and the endless competition from "me too" compounds from rival companies often discourage the attempt. Independent phytomedicine companies have been very strong in Germany. However, by the end of 1995, eleven (almost all) had been acquired by the multinational pharmaceutical firms, acknowledging the lay public's growing demand for phytomedicines in the Western world.

The business of dietary supplements in the Western World has expanded from the health store to the pharmacy. Alternative medicine includes plant-based products. Appropriate measures to ensure the quality, safety and efficacy of these either already exist or are being answered by greater legislative control by such bodies as the Food and Drug Administration of the USA and the recently created European Agency for the Evaluation of Medicinal Products, based in London.

In the USA, the Dietary Supplement and Health Education Act of 1994 recognised the class of phytotherapeutic agents derived from medicinal and aromatic plants. Furthermore, under public pressure, the US Congress set up an Office of Alternative Medicine and this office in 1994 assisted the filing of several Investigational New Drug (IND) applications, required for clinical trials of some Chinese herbal preparations. The significance of these applications was that each Chinese preparation involved several plants and yet was handled as a *single* IND. A demonstration of the contribution to efficacy, of *each* ingredient of *each* plant, was not required. This was a major step forward towards more sensible regulations in regard to phytomedicines.

My thanks are due to the staff of Taylor & Francis who have made this series possible and especially to the volume editors and their chapter contributors for the authoritative information.

Roland Hardman

Preface

Plants play a vital role in traditional medicine systems. Especially in the Far East, several medicinal plants have been used in traditional medicine preparations for centuries, and many of these preparations are still in use today. A significant number of these plants have also been exposed to modern drug discovery screening, and as a result, several modern drugs have been developed from the compounds isolated from plant sources. The genus *Magnolia* L. (Family Magnoliaceae Juss.) consists of several medicinally important species, most of them from Asia, especially the Far East. Many species of this genus have traditionally been used, largely in China and Japan, to treat various illnesses ranging from simple headaches to complicated cancers. For example, *M. grandiflora* is used to treat common cold, headaches and stomach aches, *M. obovata* is used as a crude drug for its digestive and analgesic properties, *M. obarata* to treat chronic hepatitis, *M. officinalis* for the treatment of anxiety, thrombotic stroke, typhoid, asthma and gastrointestinal complaints, *M. biondii* to cure nasal empyema and headaches, *M. virginiana* to combat malaria, and *M. hypoleuca* against cancer. While mainly the bark of these plants are used in medicine, other parts have also been reported to possess various pharmacological properties.

The genus *Magnolia* plays an important role in the Chinese and Japanese traditional medicine system. Owing to its versatile medicinal properties, several species from this genus have been incorporated and used in different commercially successful medicine preparations. Hsiao-cheng-chi-tang, Wuu-Ji-San, Heii-san, Shimpi-to, Hangekouboku-to, Masinin-gan, Sai-boku-to, Syosaiko-to, Irei-to, Goshaku-san are just a few of them.

In recent years, with the revival of the popularity of, and renewed interest in, herbal and oriental traditional medicines, many of these *Magnolia*-containing Chinese-Japanese traditional medicine preparations have captured a significant proportion of the drug market in Western countries. Moreover, the market for medicinal plants as a whole is ever-growing, and *Magnolia* certainly is one of the commercially important genera.

To date, several phytochemical, pharmacological and toxicological studies, resulting in the isolation of a number of bioactive compounds (honokiol and magnolol are the most cited ones) and discovery of new biological and pharmacological activities (anti-HIV, anti-cancer and anti-tumour, anti-convulsant, anti-microbial, anti-inflammatory, anti-emetic, anti-asthma, etc.), have been performed with *Magnolia* species or *Magnolia*-containing preparations. Aiming at the improved preparation and quality control of *Magnolia* crude drugs, several modern extraction and chromatographic methods have also been reported. The time has come for the success story of *Magnolia* to be told and documented in the form of a book, which will be a handy reference for academics and researchers and also for people involved in the *Magnolia* trade in one form or another.

Although a book by D.J. Callaway, covering predominantly biological and horticultural discussions of the genus *Magnolia*, was published in 1994, the need for a more comprehensive book detailing several other aspects, e.g. phytochemistry, pharmacology, toxicology, quality control, commercial significance, and so on remains valid. Our proposed volume on the genus *Magnolia* will certainly fulfil that need.

In closing, we would like to thank all the authors who have kindly spent their valuable time in preparing the different chapters for this book.

Dr Satyajit D. Sarker
Professor Yuji Maruyama

Contributors

Amanda Harris BSc
Solgar R&D, 1211 Sherwood Avenue, Richmond, VA 23220, USA

William A Hoch PhD
Department of Horticulture, University of Wisconsin-Madison, 312 Horticulture, 1575 Linden Dr. Madison, WI 53706-1590, USA

Syunji Horie PhD
Faculty of Pharmaceutical Sciences, Chiba University, Laboratory of Chemical Pharmacology, 1-33, Yayoi-cho, Inage-ku, Chiba-shi, Chiba 263-8522, Japan

Yasushi Ikarashi PhD
Department of Neuropsychopharmacology (Tsumura), Gunma University School of Medicine, 3-39-22 Showa-machi, Maebashi-shi, Gunma 371-8511, Japan

Fumio Ikegami PhD
Faculty of Pharmaceutical Sciences, Chiba University, Medicinal Plant Gardens, 1-33, Yayoi-cho, Inage-ku, Chiba-shi, Chiba 263-8522, Japan

Hisashi Kuribara PhD
Wakanyaku Medical Institute Ltd., Laboratory of Development, 1193 Akagiyama, Fujimimura, Seta-gun, Gunma 371-0101,

Zahid Latif MPharmS, PhD
MolecularNature Limited, Plas Gogerddan, Aberystwyth, Ceredigion SY23 3EB, Wales, UK

Yuji Maruyama PhD
Department of Neuropsychopharmacology (Tsumura), Gunma University School of Medicine, 3-39-22 Showa-machi, Maebashi-shi, Gunma 371-8511, Japan

Lutfun Nahar BSc (Hons), GRSC
Department of Chemistry, University of Aberdeen, Meston Walk, Old Aberdeen, Aberdeen AB24 3UE, Scotland, UK

Satyajit D Sarker BPharm (Hons), MPharm, PhD
School of Pharmacy, The Robert Gordon University, Schoolhill, Aberdeen AB10 1FR, Scotland, UK

Motokichi Satake PhD
Pharmacognosy and Phytochemistry Division, National Institute of Health Science, 1-18-1 Kami-Yoga, Setagaya-ku, 158-8501 Tokyo, Japan

Yu Shao PhD
Solgar R&D, 1211 Sherwood Avenue, Richmond, VA 23220, USA

Suhua Shi PhD
Key Laboratory of Gene Engineering of Ministry of Education, School of Life Sciences, Zhongshan University, Xinguang Road 135, Guangzhou 510275, China

Michael Stewart MSc (Hons), PhD
MolecularNature Limited, Plas Gogerddan, Aberystwyth, Ceredigion SY23 3EB, Wales, UK

Kazuo Toriizuka PhD
The Research Division of Oriental Medicine Research Centre, The Kitasato Institute, 5-9-1 Shirokane, Minato-ku, Tokyo 108-8642, Japan

Kazuo Watanabe PhD
Faculty of Pharmaceutical Sciences, Chiba University, Laboratory of Chemical Pharmacology, 1-33, Yayoi-cho, Inage-ku, Chiba-shi, Chiba 263-8522, Japan

Peter Zhang PhD
Nutratech, Inc., 208 Passaic Ave, Fairfield, NJ 07004, USA

Yang Zhong PhD
Key Laboratory of Biodiversity Science and Ecological Engineering, Ministry of Education, School of Life Sciences, Fudan University, Shanghai 200433, China

1 Introduction—The Genus *Magnolia*

Kazuo Watanabe, Fumio Ikegami and Syunji Horie

1.1 The Magnoliaceae family

The Magnoliaceae is a family of about 220 species of deciduous or evergreen trees and shrubs native to Asia and America, with large showy flowers containing both male and female parts. The family was named in honor of Pierre Magnol, a professor of botany and medicine at Montpelier (1638–1715). Its best-known representatives are the horticulturally important species of the genus *Magnolia*.

Approximately 80% of the species are distributed in temperate and tropical Southeast Asia from Himalayas eastward to Japan and southeastward through the Malay Archipelago to New Guiana and New Britain. The remaining 20% are found in America, from temperate southeast North America through tropical America to Brazil. All the American species belong to the three genera *Magnolia*, *Talauma* and *Liriodendron*, which occur also in Asia and thus have independent discontinuous distributions. Fossil records indicate that the family was formerly much more widely distributed in the Northern Hemisphere, e.g. in Greenland and Europe (Heywood, 1979).

1.2 Botanical features

The family to which the genus *Magnolia* belongs is morphologically very distinct and easily recognizable.

Leaves: The leaves are alternate, simple, petiolate, with large stipules, which at first surround the stem, but fall off as the leaf expands and leave a characteristic scar around the node.

Flowers: The flowers are bisexual (rarely unisexual), often large and showy, pedunculate, solitary at the ends of branches or in the axils of the leaves (sometimes paired when axillary); the penduncle bears one or more spathaceous bracts which enclose the young flower, but fall off as it expands. The perianth is composed of two or more (usually three) whorls of free tepals, which are petaloid; the outer tepals are sometimes reduced and sepal-like. The stamens are numerous, spirally arranged, with stout filaments; the anthers have two locules opening by longitudinal slits. The carpels are numerous or few (rarely single), spirally arranged, free or partly fused. Each carpel has two or more ventrally placed ovules.

Fruit: The fruit is composed of separate or united carpels, which are longitudinally dehiscent or circumscissidal or indehiscent.

Seeds: The seeds are large (except in *Liriodendron*), with an arilloid testa free from the endocarp, but attached by a silky thread in *Liriodendron* they adhere to the endocarp, without arilloid testa. The seeds have copious endosperm and a minute embryo (Heywood, 1979).

1.3   Classification

The Magnoliaceae is arguably regarded as the most primitive living family of flowering plants. This family is a unique group distinct from any other plant family. The Magnoliaceae belongs to the order Magnoliales, which comprises a number of plant families including the Winteraceae and Annonaceae. The families Winteraceae and Annonaceae have been traditionally associated with Magnoliaceae because of their similarities in floral structure.

There are 12 genera, forming two tribes in the Magnoliaceae. The genus *Liriodendron*, which is different from the others in its characteristic lobed leaves, extrose anthers and deciduous winged indehiscent fruiting capels, forms the small tribe Liriodendrae. All the other genera belong to the tribe Magnoliae, with inward or dehiscing anthers and seeds with arilloid testa, which is free from the endocarp.

The Magnoliae with terminal flowers includes the two largest genera, *Magnolia* (over 80 species) and *Tarauma* (over 50 species), both of which are common to Asia and America, as well as the Asian genera *Manglietia* (25 species), *Alcimandra* (1 species), *Aromadendron* (4 species), *Pachylarnax* (2 species) and *Kmeria* (2 species). The genera with axillary flowers are all from Asia and include the genera *Michelia* (about 40 species), *Elmerrillia* (6 species), *Paramichelia* (3 species) and *Tsoongiodendron* (1 species).

In Japan, six species of *Magnolia* and one species of *Michelia* grow in nature. The following *Magnolia* plants are popular, including horticultural ones: *Magnolia acuminata*, *M. grandiflora*, *M. heptapeta*, *M. hypoleuca* (= *M. obovata* Thunb.), *M. macrophylla*, *M. praecocissima* (= *M. kobus* DC.), *M. quinquepeta*, *M. salicifolia*, *M. sieboldii*, *M. soulangiana*, *M. thompsoniana*, *M. tomentosa*, *M. virginiana*, *M. wieseneri* and *M. sayonara* (Phillips, 1978; Kawano, 1988).

1.4   Pollination ecology in Magnoliaceae plants

Magnoliaceae plants are specialized in terms of pollination ecology. They often have a strong odor and are mainly beetle-pollinated. In some cases flowers are adopted to attract beetles. Examples are the stigmatic warts in some *Magnolia* species from which the beetles or other insects can obtain sugar-containing liquids after biting. In many cases the flowers can act as traps for beetles; the inner tepals can keep the beetle captive in their closed position. The beetles stay overnight in the flowers. Most Magnoliaceae are protogynous and when beetles are caught in the female phase they will leave the flower the next day in the male phase after pollen shedding, and the conditions for successful cross-pollination are assured. Although there are many other styles of pollination by insects, the flowers of *Magnolia* can be regarded as having been highly specialized beetle flowers since their appearance in the fossil record (Brill, 1987).

1.5 *Magnolia obovata* and *Magnolia officinalis* in Chino-Japanese traditional medicine

Among *Magnolia* species, *M. obovata* and *M. officinalis* are very important in traditional Chino-Japanese herbal medicine. In the Chinese Pharmacopoeia, there are three entries containing *Magnolia* species: Cortex Magnoliae, "Houpo" (the dry bark of stem, branch and root of *M. officinalis* and *M. officinalis* var. *biloba* Rehd. et Wils); Flos Magnoliae officinalis (the dry flower buds of *M. officinalis* or var. *biloba*); and Flos Magnoliae "Xinyi" (the dry flower buds of *Magnolia biondii* Pamp., *M. denudata* Desr., *M. sprengeri* Pamp.). The Japanese Pharmacopoeia also contains similar crude drugs other than Flos Magnoliae officinalis. The Japanese Magnoliae Cortex is the dry bark of *Magnolia obovata* Thunberg.

Chinese Magnoliae Cortex: The tree of *M. officinalis* is cultivated in the upper Yangtse provinces in China for its bark, which is called "Houpo" in Chinese and extensively used as a crude medicine. The tree has very large leaves, and large white flowers. The crude drug is the rough, thick, brown bark rolled into large, tight cylinders from 180 to 230 mm long (Figure 1.1e). The outer surface is of a grayish-brown color, roughened with tubercles and marked with lichenous growths. The inner surface is smooth and of reddish-brown color. In the coast provinces there seems to be some confusion with regard to the origin of the crude drug, which is probably the bark of a different tree also appearing in the market. There is some confusion of Chinese terms between this and *Celtis sinensis*. The taste of the true bark is aromatic and bitter, but some of the drug in the market is said to be almost tasteless, and is probably inert.

The traditional medicinal properties of *Magnolia* bark are deobstruent, tonic, stomachic, quieting, and anthelmintic uses. It is prescribed in diarrhea, flatulence, amenorrhea, pyrosis and a variety of related gastrointestinal difficulties (Stuart, 1969).

Japanese Magnoliae Cortex: *Magnolia* bark in the Japanese Pharmacopoeia is of *Magnolia obovata* Thunb. (Figure 1.1f). It has been used as a substitute for the imported expensive Chinese "Houpo" for over a thousand years in Japan. In ancient Japanese textbooks on traditional medicine, it was documented that the therapeutic activity of the "Houpo" imported from China was much superior to the Japanese substitute (Figure 1.2).

Figure 1.1 Photographs of genus *Magnolia* and *Magnolia* barks used in China and Japan as the traditional herbal medicines. (a) *Magnolia obovata* grown in Japan. (b) Flower of *M. obovata*. (c) *Magnolia denudata* grown in Japan. (d) Flower of *M. denudata*. (e) The bark of *M. officinalis* from China. (f) The bark of *M. obovata*.

1.6   Chemical constituents in *Magnolia* plants

The genus *Magnolia* is a rich source of several biologically active compounds. Phytochemical investigations on several species of *Magnolia*—*M. officinalis*, *M. obovata* and *M. biondii* in particular—have been performed to date. The chemical constituents of *M. officinalis* are most extensively studied among the Magnoliaceae plants because it is one of the important plant medicines in Chino-Japanese traditional medicine (Tang and Eisenbrand, 1975). Both the prescriptions Hange-koboku-to and Sai-boku-to in

Figure 1.2 A painting of *Magnolia obovata* Thunberg from an old Japanese textbook of medicinal plants. ("Honzozufu" painted about a hundred years ago in Japan.)

Japanese are still in use in modern clinical practice in Japan. The bark of *M. officinalis* contains three groups of biologically active compounds, alkaloids, essential oils and biphenols (e.g. magnolol and honokiol). Among them, the biphenol compounds have attracted the strong interest of pharmacologists because of their various pharmacological activities. The biphenol compounds magnolol and honokiol have been isolated from the bark of both *M. officinalis* and *M. obovata*. Quantitative determination of magnolol and honokiol by various analytical methods showed magnolol contents of 2–11% and honokiol contents of 0.3–4.6% in the stem bark. The root bark contains more magnolol and honokiol than the stem bark. Magnolol and honokiol can be extracted from the bark with methanol or sodium hydroxide solution, precipitated with hydrochloric acid, and purified by recrystallization. The quaternary ammonium compounds of magnocurarine and salicifoline were also isolated from *M. obovata* and have distinct curare-like neuromuscular blocking action. A separate chapter in this book has been devoted to this aspect of *Magnolia* (Chapter 3, Sarker *et al.*).

1.7   Scientific evaluation of *Magnolia* bark based on the traditional medicinal use

According to primitive ancient documents on traditional medicine, the causes of diseases were recognized being composed of three elemental disorders or obstructions of "Chi" (neurohumoral activity or energy), "Hsieh" (blood) and "Shui" (body fluid) (Otsuka, 1976). The role of *Magnolia* bark in the crude drug prescriptions may correspond to deobstruction of "Chi". Thus *Magnolia* bark was called a Chi-drug. We can

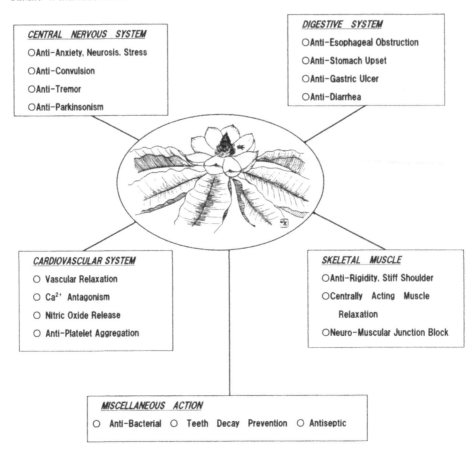

Figure 1.3 Pharmacological properties of *Magnolia* bark.

interpret the traditional medical use of *Magnolia* bark as sedative, anticonvulsive, skeletal muscle relaxing, as well as giving relief of gastrointestinal problems caused by mental stress. The present author and his group (Watanabe *et al.*, 1973, 1975, 1983a,b) found a distinct sedative and centrally acting muscle relaxant action of the ether extract from *Magnolia* bark, as well as an anti-ulcer effect in ulcers induced by stressful treatment in experimental animals (Figure 1.3). After related pharmacological research on these actions, we obtained the active principles of magnolol and honokiol, and recognized the significance of careful examination of ancient references on the basis of modern science. Recently Maruyama and his group have been extending the neuroscientific research on the active constituents in *Magnolia* bark (Maruyama and Kuribara, 2000). They have obtained promising results showing remarkable anxiolytic effects of magnolol, honokiol and related compounds. More information on the bioactivity and pharmacological properties of *Magnolia* plants is presented in Chapter 4.

1.8 Conclusion

The pharmacological and medicinal potentials of *Magnolia* plants are well documented; compilation of this book can certainly be justified to convey this message to the scientific community. We believe this book will contribute to promoting the progress of science in the field of herbal medicine associated with traditional cultures around the world.

References

Brill, E.J. (1987) *An Evolutionary Basis for Pollination Ecology*. Leiden Botanical Series, vol. 10, pp. 245–248. Leiden: Leiden University Press.

Heywood, V.H. (1979) *Flowering Plants of the World*, pp. 27–28. Oxford: Oxford University Press.

Kawano, S. (1988) *The Plant World: Newton Special Issue*, pp. 22–53. Tokyo: Kyoikusha.

Maruyama, Y. and Kuribara, H. (2000) Overview of the pharmacological features of honokiol. *CNS Drug Rev.*, 6, 35–44.

Otsuka, K. (1976) *KANPO: Geschichite, Theorie und Praxis der Chinesisch–Japanischen Traditionellen Medizin*, pp. 42–75. Tokyo: Tsumura Juntendo AG.

Phillips, R. (1978) *Trees in Britain, Europe and North America*, pp. 135–137. London.

Stuart, G.A. (1969) *Chinese Materia Medica*, 2nd edn, pp. 254–256. Taipei: Ku T'ing Book House.

Tang, W. and Eisenbrand, G. (1975) *Chinese Drugs of Plant Origin*, pp. 639–646. London: Springer-Verlag.

Watanabe, H., Watanabe, K. and Kikuti, T. (1983a) Effect of *d*-coclaurine and *d*-reticuline, benzyltetrahydroisoquinoline alkaloids, on levels of 3,4-dihydroxyphenyl acetic acid and homovanillic acid in the mouse striatum. *J. Pharmaco-bio Dyn.*, 6, 793–796.

Watanabe, K., Goto, Y. and Yoshitomi, K. (1973) Central depressant action of *Magnolia* cortex. *Chem. Pharm. Bull.*, 21, 1700–1708.

Watanabe, K., Watanabe, H., Goto, Y. and Yoshizaki, M. (1975) Studies on the active principles of magnolia bark. Centrally acting muscle relaxant activity of magnolol and honokiol. *Jpn. J. Pharmacol.*, 25, 605–607.

Watanabe, K., Watanabe, H., Goto, Y., Yamaguchi, M., Yamamoto, N. and Hagino, K. (1983b) Pharmacological properties of magnolol and honokiol, neolignane derivatives, extracted from *Magnolia officinalis*: central depressant effects. *Planta Med.*, 49, 103–108.

2 Traditional Medicine and *Magnolia*

Kazuo Toriizuka

2.1 Introduction

The theoretical foundations of traditional medicine in far-east Asia are based on and developed from medical classics such as "The Yellow Emperor's Classic of Internal Medicine" (Huang Ti Nei Ching) written in the first century BC, and "The Treatise on Febrile Diseases" (Shang Han Lun) and "The Prescriptions from the Golden Chamber" (Chin Kuei Yao Lueh), of ca. AD 218. The principles of the therapeutic methods in these texts originated from the knowledge and experience gained through clinical practice of many years. The origin of Japanese traditional medicine (Kampo medicine) can be found in these ancient Chinese traditional medical texts. However, Kampo medicine today is somewhat different from Chinese traditional medicine because of differences in historical developments, location, and national habits. One of the important aspects of these differences is the source and quality of herbs used (Hong-Yen Hsu, 1987).

"Sheng Nong Ben Cao Jing" (Shen Nong's Herbal), the earliest Chinese Materia Medica, is believed to be a product of the first century BC. In this text 365 different kinds of crude drugs were listed and categorized into three classes: superior, common and inferior. *Magnolia* bark (Magnoliae Cortex, family Magnoliaceae; Koboku in Japanese, Houpo in Chinese) is one of the crude drugs listed in the common class. It is incorporated in a number of traditional herbal preparations that are used for the treatment of gastrointestinal distension, vomiting, diarrhea, cough, and phlegm.

According to the *Dictionary of Herbal Drugs* (Chung Yao Ta Tsu Tien, 1997), the source of *Magnolia* bark (Magnoliae Cortex) should be *Magnolia officinalis* Rehder et Wilson, or *M. officinalis* var. *biloba* Rehder et Wilson (Chinese *Magnolia*; Kara-Koboku). As these plants are not available in Japan, Japanese *Magnolia* bark consists of the dried barks of *Magnolia obovata* Thunberg (Japanese *Magnolia*; Wa-Koboku) (Table 2.1). The 14th edition of the Japanese Pharmacopoeia (JPXXIV, 2001) presents both Chinese *Magnolia* and Japanese *Magnolia* as "Magnolia Bark".

Table 2.1 Modern definition of *Magnolia* bark

Magnoliaceae	People's Republic of China[a]	Japan[b]
Magnolia bark; Magnoliae Cortex	*Magnolia officinalis* Rehder et Wilson *Magnolia officinalis* var. *biloba* Rehder et Wilson	*Magnolia officinalis* Rehder et Wilson *Magnolia officinalis* var. *biloba* Rehder et Wilson *Magnolia obovata* Thunberg
Houpo (Chinese) Koboku (Japanese)	Chinese Magnolia; "Kara-koboku"	Japanese Magnolia; "Wa-koboku"

Information obtained from [a]Chung Yao Ta Tsu Tien (1997); [b]JPXIV (2001).

At the time when the ancient traditional medical texts of China were first introduced in Japan, Japanese physicians (medical doctors and pharmacists) focused on investigating whether the Chinese plants mentioned in those texts were also growing in Japan, whether they contained the same quantity of active ingredients, and/or whether they had the same pharmacological activities. One of those plants of interest is *Magnolia* bark (Koboku). The *Magnolia* bark grown in Japan used to be called "Wa-koboku" (*Wa* means Japanese) and it replaced the Chinese *Magnolia* bark "Kara-koboku" (*Kara* means Chinese) when the Chinese *Magnolia* bark was not available in Japan. Historically, it was believed that the Chinese *Magnolia* bark was of greater efficacy than the Japanese *Magnolia* bark. This chapter reviews the history and traditional medicinal uses as well as the efficacy of *Magnolia* bark in today's clinical practice.

2.2 Traditional understanding of *Magnolia* bark (Koboku)

In Kampo medicine, *Magnolia* bark is believed to be a drug that replenishes *KI* or *Chi* (meaning vital energy, the basic function for existence). The common belief is that this plant can strengthen a patient's life energy by activating circulation of vital energy and eliminating dampness. Other claimed medicinal properties of *Magnolia* include diuretic and antitussive effects. *Magnolia* bark is used for pain in the abdomen due to entrapped gas and a feeling of congestion in the chest. It is also used in psychological disorders. A summary of the traditional understanding of *Magnolia* bark is presented in Table 2.2.

The healing qualities of *Magnolia* bark can be found in "Sheng Nong Ben Cao Jing" (later Han-Dynasty about 220–25 BC). *Magnolia* bark has been described to be effective in curing headache and chill preliminary to a rise in body-temperature due to invasion of pathologic *KI* or *Chi* in the form of "coldness or wind", and against fever and delirium due to high body temperature. It has also been reported that it may be beneficial to patients suffering from paralysis and hypoesthesia, and that it may have antibacterial and antiparasitic activity.

"Ming Yi Bie Lu" (*Transactions of Famous Physicians*, a book on pharmacognosy, compiled by Tao Hong-jing in AD 500) describes *Magnolia* bark as being effective against abdominal pain and congestion, diarrhea, cough, phlegm, gonorrhea, congestion of the chest, and gastric and intestinal catarrh. It is noted that unlike "Sheng Nong Ben Cao Jing", "Ming Yi Bie Lu" cites the efficacy of *Magnolia* bark in gastric and intestinal troubles.

Table 2.2 Traditional understanding of *Magnolia* bark (Koboku)

Medical classic textbooks	Year of publication	Description	Country
Sheng Nong Ben Cao Jing (Shen Nong's Herbal)	220–225 BC	For headache and chill. Effective against fever and delirium. Beneficial to patients suffering from paralysis and hypoesthesia	China
Ming Yi Bie Lu (Transactions of Famous Physicians)	AD 500?	Effective against abdominal pain and congestion, diarrhea, gonorrhea, chest congestion, gastric and intestinal catarrh. To warm the middle part of the body. To dissolve phlegm and stop coughing.	China
Yao Xing Lun (A book on pharmacological nature of drugs)	AD 627–649	For stomach and intestinal troubles. Improves microcirculation in small blood vessels. Diuretic effect. To warm the middle part of the body.	China
Tang Ye Ben Cao (Materia Medica of Decoction)	AD 1289	Effective in diseases concerning the lungs and the bronchial system	China
Wago-honzo-komoku (The translation of the original Chinese Materia Medica into Japanese)	17th century	For stomach and intestinal troubles due to invasion of cold and/or stagnancy of fluids	Japan
Yaku-cho (A book on the theory of herbal pharmacology)	1771	To relieve congestion in the chest and abdomen	Japan
Jyu-kou-yaku-cho (A book on the theory of herbal pharmacology, additional version)	19th century	To relieve congestion in the chest and abdomen. Effective in acute and chronic cough.	Japan
Koho-yakugi (A description of drugs used by the Koho-ha school)	1863	To relieve congestion of the stomach and intestine area	Japan
Chung Yao Ta Tsu Tien	1997	To eliminate damp and relieve distension. Accumulation of damp in the spleen and the stomach marked by epigastric stuffiness, vomiting and diarrhea. Abdominal distension. Cough and dyspnea caused by retained damp.	China

"Yao Xing Lun", a book on the pharmacological nature of drugs, mentions that Koboku can deal with gastric and intestinal troubles that are due to pathological conditions associated with cold and symptoms of stagnation of food. *Magnolia* bark also improves microcirculation in small blood vessels and has a diuretic effect. Because of its ability to warm the middle part of the body (the ancient terminology referring to the triple warmer of acupuncture), it reduces pain in this area, has anti-emetic properties and dispels stagnant water in the stomach. In the book "Tang Ye Ben Cao" (*Materia Medica of Decoction*, written by Wang Hao-gu in AD 1289), the efficacy of

Magnolia bark in the treatment of diseases concerning the lungs and bronchial systems is mentioned.

In Japan, authors like Todo Yoshimasu (1702–1773) in his book "Yaku-cho" (a book on theoretical herbal pharmacology, 1771) mentioned that *Magnolia* bark is effective in all diseases with symptoms of congestion in the chest and abdomen, In addition to the properties mentioned by Yashimasu, Yodo Odai (1799–1870), in his book "Jyu-kou-yaku-cho" (on theoretical herbal pharmacology, additional version, nineteenth century) described *Magnolia* bark as effective in the treatment of acute or chronic cough. In the book "Wago-honzo-komoku" (the interpretation of the original Chinese Materia Medica into Japanese, seventeenth century), Ippo Okamoto (about 1654–1716) stressed the efficacy of *Magnolia* bark in gastrointestinal disturbances associated with invasion of cold and/or stagnation of fluids. In "Koho-Yakugi" (a description of drugs used by the Koho-ha school, 1863), Sohaku Asada (1815–1894) pointed out that many symptoms in other parts of the body have their origin in congestion of the stomach and the intestinal area, and that *Magnolia* bark is an excellent remedy for these ailments.

In ancient books such as "Sheng Nong Ben Cao Jing", the main effects of *Magnolia* bark were mentioned as anti-palsy and sedative, whereas according to the books published later than "Ming Yi Bie Lu" the efficacy was reported to be in illness concerning the digestive system. Books published after "Tang Ye Ben Cao" deal mainly with the effect of *Magnolia* bark on respiratory problems. The effects of *Magnolia* bark are summarized below:

- Removes congestive feelings in abdomen and chest
- Strengthens the stomach
- Dispels abdominal pain
- Sedative effect by inhibiting the rise of *KI* or *Chi*
- Antitussive effect
- Mucolytic effect
- Improves microcirculation of blood

In the "Dictionary of Herbal Drugs" a similar description of the activity of *Magnolia* bark can be found as follows:

- Eliminates damp and relieves distension
- Relieves accumulation of damp in the spleen and the stomach marked by epigastric stuffiness, vomiting and diarrhea
- Relieves abdominal distension due to retention of undigested food
- Cures cough and dyspnea caused by retained damp

2.3 *Magnolia* bark (Koboku) in Kampo prescriptions

The foundation of Kampo prescriptions is based on the information gathered over a long period of clinical practice and accumulation of empirical data, whereas research, in the modern pharmaceutical sense, has only a short history of some decades (Toriizuka, 2000). Yamazaki (1986) has listed Kampo prescriptions using *Magnolia* bark as a component, referring to the classical books of Kampo medicine such as

"Shang Han Lun" (*Treatise on Febrile Diseases*, a medical text written by Zhang Zhong-jing about AD 200), "Jin Kui Yao Lue" (a section of the "Shang Han Lun" devoted mainly to the treatment of chronic diseases, third century), and "Futsugo-Yakuhitu-Hokan" (teaching notes on the application of formulations of Sohaku Asada (ca. 1815–1894) (see Tables 2.3 and 2.4). A summary of these prescriptions is presented below.

2.3.1   "Shang Han Lun" and "Jin Kui Yao Lue"

Group I: Prescriptions against congestion of the chest and abdomen

Dai-joki-to (Da-cheng-qi-tang; Formula animationis major) is an example of this type of prescription. It consists of four components, Magnoliae Cortex, Rhei Rhizoma, Natrium sulfate, and Aurantii Fructus immaturus. Dai-joki-to is normally prescibed for treating congestion and pain in the abdomen, and congestion of the chest plus pain and dyspnea. It was also used for encephalitis and poliomyelitis. The prescription Shojoki-to (Xia-cheng-tang; Formula animationis minor) consists of three components, Magnoliae Cortex, Rhei Rhizoma and Aurantii Fructus immaturus, and is prescribed for the treatment of constipation, meteorism and delirium due to high body temperature. Koboku-sanmotsu-to (Houpo-san-wu-tang; Formula magnoliae) is a prescription similar to Shojoki-to, but it uses a higher dosage of Magnoliae Cortex and Aurantii Fructus immaturus for patients with aggravated symptoms of constipation and abdominal pain. This type of prescription is therefore composed of drugs with purgative action like Rhei Rhizoma, and Natrium sulfate together with herbs that can cure the symptoms of loss of appetite and act as antiemetic agent.

Group II: Prescriptions against cough and asthma

The "Sheng Nong Ben Cao Jing" and the "Ming Yi Bie Lu" do not mention the effectiveness of *Magnolia* bark against cough and asthma. However, Keishi-ka-koboku-kyonin-to (Gui-zhi-jia-houpo-xing-ren-tang; Formula cinnamonic cum magnoliae e t armaniacae) and Koboku-mao-to (Houpo-ma-huang-tang; Formula magnoliae et ephedrae), listed in the "Shang Han Lun" and the "Jin Kui Yao Lue", are two preparations that are prescribed to treat cough, asthma, and other diseases associated with the respiratory system. In these prescriptions *Magnolia* bark functions as a *KI/Chi*-lowering drug and therefore improves respiratory conditions. It is assumed that *Magnolia* bark together with the other components of the prescriptions is most effective against symptoms aggravated by emotional stress. In addition to these preparations, Saiboku-to (Chai-po-tang; Formula magnoliae et bupleuri) is proven (by clinical and other scientific research data) to be effective in the treatment of asthma.

Group III: Prescriptions against pain in the chest and abdomen

The *Magnolia* bark-containing preparation Kijitsu-gaihaku-keishi-to (Zhi-shi-xie-bai-gui-zhi-tang; Formula aurantii allii cinnamomi) is used for relieving pain in the chest and abdomen. Nowadays, this preparation is frequently prescribed for the treatment of stenocardia and myocardial infarction.

Table 2.3 Kampo prescriptions containing *Magnolia bark* (I) (summarized from Shang Han Lun and Jin Kui Yao Lue)

Prescriptions	Magnoliae Cortex	Aurantii Fructus Immaturus	Rhei Rhizoma	Armeniacae Semen	Pinelliae Tuber	Zingiberis Rhizoma	Glycyrrhizae Radix	Zizyphi Fructus	Cinnamomi Cortex	Hoelen	Perillae Herba	Others
Group I: Prescriptions against congestion of the chest and abdomen												
Dai-joki-to (Da-cheng-qi-tang; Formula animationis major)	*	*	*									Natrii Sulfus
Shojoki-to (Xia-cheng-tang; Formula animationis minor)	*	*	*									
Koboku-sanmotsu-to (Hou-po-san-wu-tang; Formula magnoliae)	*	*	*									
Group II: Prescriptions against cough and asthma												
Keishi-ka-koboku-to (Gui-zhi-jia-houpo-xing-ren-tang; Formula cinnamomic cum magnoliae et armaniacae)	*			*		*	*	*	*			Paeoniae Radix
Koboku-mao-to (Houpo-ma-huang-tang; Formula magnoliae et ephedrae)	*			*	*	*[a]						Asiasari Radix, Ephedrae Herba, Gypsum Fibrosum, Schisandrae Fructus, Tritici Semen
Saiboku-to (Chai-po-tang; Formula magnoliae et bupleuri)	*				*	*	*	*		*	*	Bupleuri Radix, Ginseng Radix, Scutellariae Radix
Group III: Prescriptions against pain in the chest and abdomen												
Kijitsu-gaihaku-keishi-to (Zhi-shi-xie-bai-gui-zhi-tang; Formula aurantii allii cinnamomi)	*	*							*			Allii Chinensis Bulbus, Trichosanthis Semen
Group IV: Prescriptions against stagnant KI/Chi												
Hange-koboku-to (Ban-xia-houpo-tang; Formula magnoliae et pinelliae)	*				*	*				*	*	

[a] = Instead of Zingiberis Rhizoma, Zingiberis Siccatum Rhizoma is used in Koboku-mao-to.

Table 2.4 Kampo prescriptions containing *Magnolia* bark (II) (summarized from Futsugo-Yakuhitu-Hokan)

Group I: For meteorism, hydrops and as a laxative agent

Prescriptions	Magnoliae Cortex	Aurantii Fructus Immaturus	Rhei Rhizoma	Cannabis Fructus	Armeniacae Semen	Saussureae Radix	Aurantii Nobilis Pericarpium	Perillae Herba	Atractylodis Lanceae Rhizoma	Atractylodis Rhizoma	Hoelen	Polyporus	Alismatis Rhizoma	Pinelliae Tuber	Astragali Radix	Ginseng Radix	Platycodi Radix	Angelicae Dahuricae Radix	Angelicae Radix	Cnidii Rhizoma	Paeoniae Radix	Cinnamomi Cortex	Zingiberis Rhizoma	Zizyphi Fructus	Glycyrrhizae Radix	Saposhnikoviae Radix	Others
Mashinin-gan (Ma-zi-ren-wan; Pilulae cannabis)	*	*	*	*	*																*						
Juncho-to (Ren-chang-tang; Formula humectandi intestini)	*	*	*	*	*														*						*		Rehmanniae Radix, Scutellariae Radix, Persicae Semen
Kumi-binryo-to (Jiu-wei-bing-lang-tang; Formula novem-arecae)	*		*			*	*	*														*	*		*		Arecae Semen
Sohakuhi-to (Sang-bai-pi-tang; Formula morus)	*							*						*		*							*		*		Mori Cortex, Anemarrhenae Rhizoma, Fritillariae Bulbus, Farfarae Flos, Schisandrae Fructus
Bunsho-to (Fen-xiao-tang; Formula atractylodis compositum)	*						*		*	*	*	*	*										*				Cyperi Rhizoma, Arecae Pericarpium, Junci Callus Medulla, Amomi Semen

Group II: For minor intestinal problems and gastric troubles

Hei-i-san (Ping-wei-san; Pulvis pacandi stomachi)

Group III: For treating diarrhea, abdominal pain and vomiting

Irei-to (Wei-ling-tang; Formula hoelen stomachi)

Ompi-to (Wen-pi-tang; Formula lienalis tepefactum) — Aconiti Tuber

Group IV: For asthma and respiratory problems

Jizen-ippo (Zhi-chuan-yi-fang; Decontumasthma)

Soshi-koki-to (Su-zi-jiang-qi-tang; Formula perillae compostae) — Peucedani Radix

Group V: For pain in the chest and abdomen

Toki-to (Dang-gui-Fructus tang; Formula angelicae) — Zanthoxyli Fructus

Goshaku-san (Wu-ji-san; Pulvis quinque-hystericorum) — Ephedrae Herba

Group VI: To dissolve stagnation of Ki/Chi in the laryngopharyngeal region

Group VII: For skin problems

Nai-taku-san (Nei-tuo-san; Pulvis antidoti)

Taku-ri-shodoku-in (Tuo-li-xiao-du-yin; Decoctum antidoti) — Gleditsiae Fructus, Trichosanthis Radix, Lonicerae Flos

Group VIII: For improving stagnation of microcirculation in peripheral blood vessels

Tsudo-san (Tong-dao-san; Pulvis purgtionis sanguinolentiae) — Akebiae Caulis, Carthami Flos, Sappan Lignum, Natrii Sulfus

Note: Instead of Zingiberis Rhizoma, Zingiberis Siccatum Rhizoma is used in the prescriptions Ompi-to, Toki-to and Goshaku-san; Aurantii Fructus Immaturus is replaced by Aurantii Fructus (mature) in the preparations Goshaku-san and Tsudo-san.

Group IV: Prescriptions against stagnant KI/Chi

The prescription Hange-koboku-to (Ban-xia-houpo-tang; Formula magnoliae et pinelliae) is effective not only in respiratory problems like Group II prescriptions, but also in the treatment of menopausal symptoms or emotional stress. The traditional belief is that this prescription dispels stagnant *KI/Chi*.

2.3.2   "Futsugo-Yakuhitu-Hokan"

The indications of medicinal uses of *Magnolia* preparations provided in the two books "Shang Han Lun" and "Jin Kui Yao Lue" are further extended in "Futsugo-Yakuhitu-Hokan", which includes stagnation of microcirculation of the blood and hydrops. However, in these preparations many components apart from *Magnolia* bark are blended, and thus it is difficult to discern whether those actions are actually due to *Magnolia* bark.

Group I

Mashinin-gan (Ma-zi-ren-wan; Pilulae cannabis) and Juncho-to (Ren-chang-tang; Formula humectandi intestini) are used for meteorism, hydrops and as laxative. Both of these preparations are used like Shojoki-to (Formula animationis minor) and Koboku-sanmotsu-to (Formula magnoliae). The main crude drug components in Mashinin-gan and Juncho-to are same as in Shojoki-to and Koboku-sanmotsu-to. For hydrops there are further prescriptions effective as Kumi-binryo-to (Jiu-wei-bing-lang-tang; Formula novem-arecae), Sohakuhi-to (Sang-bai-pi-tang; Formula morus) and others. Bunsho-to (Fen-xiao-tang; Formula atractylodis compositum) is prescribed for patients with edema due to cirrhosis of the liver and nephrosis.

Group II

For not-too-severe intestinal problems and gastric disturbances, such as stagnation of food in the stomach or enhanced peristalsis, the prescription Heii-san (Ping-wei-san; Pulvis pacandi stomachi) may be mentioned as an example. It is also effective in mild gastritis.

Group III

For treating diarrhea, abdominal pain and vomiting, Irei-to (Wei-ling-tang; Formula hoelen stomachi) and Ompi-to (Wen-pi-tang; Formula lenalis tepefactum) are widely used.

Group IV

The prescription containing Poria cocos Wolff, Armeniacae Semen, Glycyrrhicae Radix, plus Magnoliae Cortex, Cinnamomi Cassiae and Perillae Semen is called Jizen-ippo (Zhi-chuan-yi-fang; Decontum asthma), which was created by Tokaku Wada (1742–1803) from Bukuryo-kyonin-kanzo-to (Fu-ling-xing-ren-gan-cao-tang; Formula hoelen cum armeniacae et glycyrrhizae) in "Jin Kui Yao Lue", is used for asthma and respiratory problems. Soshi-koki-to (Su-zi-jiang-qi-tang; Formula perillae compostae) is used in acute and chronic asthma with more or less severe dyspnea.

Group V

Toki-to (Dang-gui-tang; Formula angelicae) is used for cold especially in the middle of the back, abdominal pain, meteorism, and it is also effective in lessening symptoms of stenocardia. Goshaku-san (Wu-ji-san; Pulvis quinque-hystericorum) is used in patients with pain in the lumbar region due to invasion of cold, cramps, feeling hot in the upper part of the body and simultaneously feeling cold in the lower part of the body, complaining of pain in the lower part of the abdomen.

Group VI

There are some preparations containing *Magnolia* bark that help to dissolve stagnation of *KI/Chi* in the laryngopharyngeal region.

Group VII

Magnolia bark is also found in prescriptions helpful for patients with skin problems. Putrid inflammation of the epidermis can be treated with Nai-taku-san (Nei-tuo-san; Pulvis antidoti) or Taku-ri-shodoku-in (Tuo-li-xiao-du-yin; Decoctum antidoti). "Nai-taku (neituo)" refers to promotion of pus discharge and a treatment for pyogenic infection of skin by oral drugs.

Group VIII

In prescriptions like Tsudo-san (Tong-dao-san; Pulvis purgtionis sanguinolentiae), *Magnolia* bark is used for improving stagnation of microcirculation in peripheral blood vessels.

2.4 Quality of *Magnolia* bark

2.4.1 *Historical studies*

Before and after the Edo period, *Magnolia* grown in Japan was used commonly by Kampo clinicians, and a lot of clinical experience was accumulated from the use Japanese *Magnolia* over a long period. However, Japanese *Magnolia* is considered to be a substitute for Chinese *Magnolia*. Nishimoto (1986) compared several medical books written in the Edo period dealing with geographic locations of the herb *Magnolia* and its varying quality. Among these books, Shuan Kagawa (1683–1755) stated, in his medical volume on drugs "Ippondo-yaku-sen (1729)", that the Chinese *Magnolia* bark (Kara-koboku) is better than the Japanese, and in fact that the Japanese plant (Wa-koboku) actually is Hou-no-ki (*Magnolia obovata* Thunberg). Its size and appearance are similar to those of the Chinese Koboku but its blossoms and fruits are different. The taste is also different. Shuan Kagawa noted that the description of *Magnolia* bark in "Ben-cao-tu-jing" (an illustrated Materia Medica edited by Su Song, 1061) clearly indicated a difference between the Chinese and Japanese *Magnolia* bark.

Todo Yoshimasu (1702–1773) in his book "Yaku-cho" (a book on the theoretical aspects of herbal pharmacology, published in 1771) wrote that there is only one Chinese *Magnolia* whereas there are two varieties of Japanese *Magnolia*, one of which is of the same quality as the Chinese one. Shoen Naito (nineteenth century) quoted in his book "Koho-Yakuhin-ko" (a description of drugs used by Kampo clinicians, published in 1841) that a *Magnolia* of high quality is that from the Satsuma area, the southern part of Japan. This indicates that, at that time, Japanese physicians began to look more intensively for a similar quality *Magnolia* bark in Japan because of the problems of importing the so called "true *Magnolia*" from China, and also because of the inferior quality of *Magnolia* bark available in Japan at that time. Sohaku Asada (1815–1894), in his book "Koho-yakugi" (published in 1863), described various *Magnolia* species. He too mentioned that Satsuma-Koboku (a name for the Japanese *Magnolia* species) is of good quality but not as good as the Chinese.

2.4.2   Studies on ancient herbal drugs stored in Shoso-in repository from the eighth century

The treasures owned by the emperor Shomu (701–756) were dedicated to Great Buddha of Todaiji-Temple by the Empress Dowager Komyo on 21 June 756. These treasures were stored in a wooden storehouse named the "Shoso-in repository" for more than 1200 years (Shibata, 2000). The collection includes valuable works of fine arts, ancient tools, musical instruments, and also crude drugs with a dedicatory record. The document, named "Shuju-yaku-cho", presents clearly-written names of crude drugs, the date and signatures of contributors certifying the authenticity of these stored drugs of the eighth century.

For more than a thousand years Shoso-in was kept closed under the Imperial Seal until the end of World War II in 1945. During this long period, none of the stored materials was offered for scientific study. The first systematic scientific investigation on Shoso-in medicines was carried out during 1948–1949. In 1994–1995, nearly 50 years after this first investigation, a second investigation was performed in collaboration with the Office of the Shoso-in Treasure House, the Imperial Household Agency, Japan. *Magnolia* bark is also stored in Shoso-in medicaments, and it is considered to be the so called "true *Magnolia* bark". In the first investigation, the morphological and microscopic features described for item N-84, which named *Magnolia* bark, could not provide any conclusive evidence to identify it in comparison with *Magnolia* bark available in the present drug market. Item N-84 is different from the present "true Chinese *Magnolia* bark", the bark of *Magnolia officinalis* Rehder et Wilson, and *M. officinalis* var. *biloba* Rehder et Wilson, since it does not have any characteristic oil cells but has a large number of clustered crystals of calcium oxalate. Some other congeners of *Magnolia* species—*M. obovata* Thunberg (*M. hypoleuca* Siebold & Zuccarini), *M. kobus* De Candolle, *M. grandiflora* Linn., *M. denudata* Desrousseaux, *M. liliflora* Desrousseaux and *M. salicifolia* Maximo also showed different features in their barks from those of item N-84.

In the second investigation, chemical examination was performed not only on *Magnolia* but also on *Michelia* and *Manglietia*. The characteristic neolignan principles, honokiol and magnolol, of the present *Magnolia officinalis* bark were found to be absent in item N-84, which has been preserved from ancient time and said to be the "true" Chinese *Magnolia* bark. However, a faint blue spot observed on TLC of extracts of item

N-84 was also observed on that of *Michelia* species extracts, especially on a terpenoid fraction of *Michelia maclureia* var. *sublanca* (Magnoliaceae). Because of this, it might be suggested that the "true Koboku" of China was not always of the same quality or origin as it was supposed to be, or that over the years degradation and decomposition of chemical constituents may occur to such an extent that they are no longer detectable.

2.5   Conclusion

According to the latest chemical and pharmacological studies on *Magnolia* bark (Magnolia Cortex; *Magnolia officinalis* Rehder et Wilson, *M. officinalis* var. *biloba* Rehder et Wilson, *M. obovata* Thunberg), the main chemical compounds and their effects have been elucidated. Magnocurarine has a curare-like effect on skeletal muscles. Magnolol and honokiol act as sedatives on the central nervous system. Magnolol prevents stress-induced gastric ulcer by inhibiting secretion of hydrochloric acid. These recent observations might support some of the clinical traditional uses of *Magnolia* bark (Koboku):

- Dispelling congestion and pain in the chest and abdomen
- Strengthening the functioning of intestine and stomach
- Sedative action according to *KI*/*Chi*-lowering activity
- Antitussive effects
- Reducing edema
- Dissolving stagnation of peripheral blood circulation in microvessels

Magnolia bark is used for a sensation of fullness in the chest and abdomen, and used as a diuretic and an antitussive in Chinese traditional medicines. The ancient descriptions of empirical observed effects find scientific explanation and support from these chemical findings and their effects on various tissues and organs in studies *in vitro* and *in vivo*. On the other hand, according to the historical, botanical and pharmacognological studies on Koboku, and chemical research on item N-84 in the Shoso-in medicaments, it has been suggested that the "true Koboku" in ancient times—at least the eighth century—was not always of the same origin or quality as was supposed to apply to *Magnolia* bark. Further scientific studies are essential for the identification of *Magnolia* bark.

Acknowledgments

The author expresses his gratitude to Dr Helmut Bacowsky for his critical reading and translation of the manuscript, and would like to thank Dr Masakazu Yamazaki and Ms Yukiko Maruyama for their helpful advice.

References

Chung Yao Ta Tsu Tien (1997) Shanghai: Shanghai Sciences and Technology Publishing.
Hong-Yen Hsu (1987) *Oriental Healing Arts Int. Bull.*, 12, 409–421.
JPXIV (2001) *Japanese Pharmacopoeia*, 14th edn. Tokyo: Hirokawa Publishing Co.
Nishimoto, K. (1986) The Quality of *Magnolia* Bark. *Journal of Traditional Sino-Japanese Medicine*, 7, 68–72. [In Japanese]

3 Phytochemistry of the Genus *Magnolia*

*Satyajit D. Sarker, Zahid Latif,
Michael Stewart and Lutfun Nahar*

3.1 Introduction
3.2 Secondary metabolites from *Magnolia*
3.3 Chemotaxonomic significance

3.1 Introduction

The genus *Magnolia* L. belongs to the family Magnoliaceae, and is named after Pierre Magnol, a distinguished professor of medicine and botany of Montpellier in the early eighteenth century. This genus comprises ca. 100 species, and most plants of this genus usually bear a large number of odoriferous flowers which are pollinated by beetles (Azuma *et al.*, 1996). Many of these plants are medicinally important and are used in traditional medicines, especially in the Far East (Namba, 1980). A number of traditional medicine preparations in China and Japan contain plant-parts of *Magnolia*, and Hsiao-cheng-chi-tang, Wuu-Ji-San, Heii-San, Shimpi-to Hangekouboku-to, Masainin, and Sai-boku-to are just a few examples of *Magnolia*-containing medicine preparations. Because of its tremendous value in traditional health-care systems, several species of this genus have been the subject of numerous phytochemical and pharmacological investigations over the last century. All these studies have been directed towards the isolation of biologically active compounds and to rationalising the use of these plants as medicines. Reports of phytochemical studies on ca. 40 *Magnolia* species are available to date. However, only about 60% of them have been investigated thoroughly. At least 255 different plant secondary metabolites, predominantly alkaloids, flavonoids, lignans, neolignans and terpenoids, have been isolated from different species of *Magnolia* (Table 3.1). Research articles published on *Magnolia* during the last century have been surveyed extensively. This chapter reviews the outcome of this survey and presents a critical discussion of several phytochemical aspects of the genus *Magnolia*.

3.2 Secondary metabolites from *Magnolia*

3.2.1 Alkaloids

Alkaloids are nitrogenous organic cyclic compounds, normally with basic properties and having physiological effects in animals or man. While alkaloids are not found in all plant families, the dicotyledonous family Magnoliaceae is well known for producing

Table 3.1 Plant secondary metabolites reported from *Magnolia* species

Name of species	Plant parts[a]	Compounds	Type	References
Magnolia acuminata L.	L	Magnoflorine (37)	Alkaloid	Kapadia *et al.* (1964a)
				Furmanowa and Jozefowicz (1980)
		Methylarmepavine (30)		Kapadia *et al.* (1964b)
		Armepavine (29)		Furmanowa and Jozefowicz (1980)
		Asimilobine (23)		
		Liriodenine (17)		
		N-Norarmepavine (31)		
		Nornuciferine (28)		
		Roemerine (22)		
	B	Magnocurarine (36)	Neolignan	Tomita and Nakano (1957)
		Acuminatin (117)	Lignan	Doskorch and Flom (1972)
		Calopiptin (94)		
		Galgravin (91)		
		Veraguensin (79)		DNP (1999)
		Machilusin (88)		
Magnolia asbei Weath.	L	Cyanidin (46)	Flavonoid	Santamour (1965b)
Magnolia biondii Pamp.	F	Buddlenoid A (53)	Flavonoid	Kubo and Yokokawa (1992)
		Biondinin A (141)	Neolignan	Ma *et al.* (1996), Pan *et al.* (1987)
	FB	Aschantin (75)		
		Biondinin B (77)	Lignan	Ma *et al.* (1996)
		Biondinin E (78)		
		Eudesmin (pinoresinol dimethyl ether) (60)		
		Fargesin (70)		
		Liroresinol A dimethyl ether (68)		
		Magnolin (64)		
		(+)-Spinescin (Kobusin) (74)		
		Yangambin (62)		
		(7R,7′m,8R,8′R)-7′,9-Dihydroxy-3,3′,4,4′-tetramethoxy-7,9′-epoxylignan (95)		
	F	Biondinin C (187)	Monoterpene	DNP (1999)
		Biondinin D (188)		Han (1993)
				DNP (1999)
Magnolia campbellii Hook.	B	Lanuginosine (14)	Alkaloid	Talpatra *et al.* (1975)
		Liriodenine (17)		
		(−)-Sesamin (73)	Lignan	

Species	Part	Compounds	Type	References
Magnolia coco (Lour.) DC.	B	Liriodenine (17)	Alkaloid	Yang *et al.* (1962)
		Magnoflorine (37)		
		Salicifoline (3)		Yang and Liu (1973)
		Anolobine (9)		
		Stephanine (24)		
	L	Magnococline (34)		Yang (1971)
		N-Acetylanolobine (8)		Yu *et al.* (1998b)
		Dicentrinone (13)		
		Oxoanolobine (16)		
Magnolia compressa Maxim.		Oxyacanthine (40)	Alkaloid	DNP (1999)
		Magnolamide (6)		Yu *et al.* (1998a)
		Scoparone (41)	Coumarin	Yu *et al.* (1998b)
		Dihydrodehydrodiconiferylalcohol (148)	Neolignan	
		Magnolol (112)	Lignan	Yu *et al.* (1998a)
		Magnolone (85)		Yu *et al.* (1998b)
		Aschantin (75)		
		Episesamin (76)		
		Eudesmin (60)		
		Epieudesmin (69)		
		Fargesin (70)		
		3-O-Demethylmagnolin (65)		
		(+)-Phillygenin (71)		
		Pinoresinol (59)		
		Sesamine (73)		
		Sesaminone (80)		
		Syringaresinol (61)		
		Syringaresinol glucoside (63)		
Magnolia cordata Michx.	FR	Peonidin 3-rutinoside (49)	Flavonoid	Santamour (1966a)
		Peonidin 3,5-diglucoside (50)		
Magnolia denudata Desr.	B	Magnocurarine (36)	Alkaloid	Tomita and Nakano (1952d)
	L			Furmanowa and Jozefowicz (1980)
	R	Magnoflorine (37)		Nakano (1956)
	L			Furmanowa and Jozefowicz (1980)
		Armepavine (29)		
		Asimilobine (23)		
		Liriodenine (17)		

Table 3.1 (cont'd)

Name of species	Plant parts[a]	Compounds	Type	References
	B	N-Norarmepavine (31)		Tomita and Nakano (1952c)
		Nornuciferine (28)		Matsutani and Shiba (1975)
		Roemerine (22)		
		Salicifoline (3)	Flavonoid	Santamour (1965b)
	L	Tyramine (4)		Santamour (1966a)
		Cyanidin (46)		
	FR	Peonidin 3-rutinoside (49)	Neolignan	Iida et al. (1982b)
	L	Burchellin (142)		
		Denudatin A (144)		
		Denudatin B (145)		
		Denudatone (152)		
		Futoenone (153)		
	FB	Licarin A (154)	Lignan	Kwon et al. (1999)
	L	Eudesmin (60)		Funayama et al. (1995)
		Pinoresinol (59)		
		Sesamin (73)		
		Spinescin (Kobusin) (74)		
		Veraguensin (79)		
		Deacyllaserine (166)	Phenylpropanoid	Iida et al. (1982b)
		trans-Isomyristicine (171)		Funayama et al. (1995)
		Sinapyl alcohol (174)		
		Costunolide (204)	Sesquiterpene	
		Parthenolide (212)		
		Myristicine aldehyde (243)	Others	Azuma et al. (1996)
		Naphthalene (254)		Funayama et al. (1995)
Magnolia fargesii Cheng	FB	Tiliroside (55)	Flavonoid	Jung et al. (1998a)
		Denudatin B (145)	Neolignan	Yu et al. (1990)
		Fargesone A (149)		Chen et al. (1988)
		Fargesone C (151)		
		Fargesone B (150)	Lignan	Teng et al. (1990)
		3-O-Demethylmagnolin (65)		Miyazawa et al. (1992)
		Eudesmin (60)		Kakisawa et al. (1972)
				Chae et al. (1998)

Species	Part	Compound	Class	Reference
		Sesamin (73)		Kakisawa et al. (1972)
		Yangambin (62)		Chae et al. (1998)
		Fargesin (70)		
		Magnolin (64)		Miyazawa et al. (1996)
		(−)-Magnofargesin (87)		Miyazawa et al. (1995)
		(+)-Magnoliadiol (97)		
		Epimagnolin A (66)		Miyazawa et al. (1994)
		Magnone A (83)		Jung et al. (1998c)
		Magnone B (84)		
		(+)-Phillygenin (71)		Miyazawa et al. (1992)
		Pinoresinol (59)		
		(−)-Fargesol (93)		
		Syringin (185)	Phenylpropanoid	Huang et al. (1990)
		Homalomenol A (225)	Sesquiterpene	Jung et al. (1998b)
		Oplodiol (220)		Jung et al. (1997)
		Oplopanone (224)		
		1β,4β,7α-Trihydroxyeudesmane (217)		
		5α,7α(H)-6,8-cycloeudesma-1β,4β-diol (219)		Jung et al. (1998b)
		Cineol (194)	Monoterpene	Kakisawa et al. (1972)
		Citral A (190)		
		α-Pinene (196)		
Magnolia fraseri Walter	F	Isoquercetin (51) Rutin (58)	Flavonoid	Santamour (1966b)
Magnolia fuscata Andrews	L	Magnoline [(−)-berbamunine] (39) Magnolamine (38)	Alkaloid	Proskurnina and Orekhov (1938) Proskurnina and Orekhov (1940) Komissarova (1945) Proskurnina (1946) Tomita et al. (1951) Tomita and Kugo (1954)
Magnolia grandiflora L.	R	Candicine (1) Salicifoline (3)	Alkaloid	Nakano (1954a), Rao (1975)
	B			
	R	Magnoflorine (37)		Nakano (1954b), Rao (1975) Nakano (1956, Tomita et al. (1961)

Table 3.1 (cont'd)

Name of species	Plant parts[a]	Compounds	Type	References
	B	Magnocurarine (36)		Rao (1975)
	W	Anolobine (9)		Tomita and Kazuka (1967)
	W & L	Anonaine (7)		
		Liriodenine (17)		
		Nornuciferine (28)		Ziyaev et al. (1999)
	L	Roemerine (22)		Matsutani and Shiba (1975)
	L & B	Tyramine (4)		Yang et al. (1994)
	L	6-Methoxy-7-hydroxycoumarin (43)	Coumarin	
		6,8-Dimethoxy-7-hydroxycoumarin (44)		
	FR	Peonidin 3-rutinoside (49)	Flavonoid	Santamour (1966a)
	S	Honokiol (108)	Neolignan	El-Feraly and Li (1978)
		Magnolol (112)		
		3,5'-Diallyl-2'-hydroxy-4-methoxybiphenyl (110)		Clark et al. (1981)
		4-O-Methylhonokiol (109)		
	B	Magnolenin C (86)		Rao and Davis (1982b)
				Rao (1975)
			Lignan	Rao and Wu (1978)
		Lirioresinol A (67)		
		Syringaresinol glucoside (73)		
		Sesamine (73)		Baures et al. (1992)
		Syringin (185)	Phenylpropanoid	Rao and Wu (1978), Juneau (1972)
				Rao (1975), Rao and Juneau (1975)
		Magnolidin (182)		
	L & ST	Parthenolide (212)	Sesquiterpene	Wiedlhopf et al. (1973)
	L	Peroxycostunolide (verlotorin) (211)		El-Feraly et al. (1977, 1979a)
		Peroxyparthenolide (214)		
	RB	Costunolide-1,10-epoxide (209)		El-Feraly et al. (1979b)
		Magnolialide (221)		
		Reynosin (223)		
		Santamarin (222)		
	B	Cycloclorenone (233)		Jacyno et al. (1991)
				Rao and Davis (1982a)

Species	Part	Compound	Type	Reference
	L	Costunolide (204)		Halim et al. (1984)
		Magnograndiolide (231)		
		Costunolide di-epoxide (210)		
		Melampomagnolide A (215)		
		Melampomagnolide B (213)		
	B	Syringaldehyde (244)	Others	El-Feraly (1984)
		Vanillin (245)		Shinoda et al. (1976)
Magnolia henryi Desr.	B	4',5-Diallyl-2-hydroxy-3-methoxydiphenyl ether (102)	Neolignan	Kijjoa et al. (1989)
		5,5'-Diallyl-2,2'-dihydroxy-3-methoxybiphenyl (111)		
		Magnolol (112)		
Magnolia kachirachirai Dandy.	B	Glaucine (19)	Alkaloid	Yang et al. (1962)
	D	Liriodenine (17)		Tomita et al. (1968)
	B	N-Norarmepavine (31)		Yang et al. (1962)
	W			Yang and Lu (1963)
	D			Tomita et al. (1968)
		Norglaucine (27)		
		Magnoflorine (37)		
		1,2,9,10-Tetramethoxy 7H-dibenzo[de,g]quinolin-7-one (18)		
	L	Acuminatin (117)	Neolignan	Ito et al. (1984b)
		Kachirachirol A (156)		Ito et al. (1984a,b)
		Kachirachirol B (157)		
		(+/−) Licarin A (154)		Li and El-Feraly (1981)
		Licarin B (155)		
		Eupomatenoid 1 (146)		Ito et al. (1984b)
		Eupomatenoid 7 (147)		
	L	(+/−) Galbacin (92)	Lignan	Ito et al. (1984a,b)
		Guaiacin (96)		
		Caryophyllene (226)	Sesquiterpene	Ito et al. (1984b)
Magnolia kobus DC.	B	Salicifoline (3)	Alkaloid	Tomita and Nakano (1952a,b)
				Nakano and Uchiyama (1956b)
	L	Magnoflorine (37)		Furmanowa and Jozefowicz (1980)
		Asimilobine (23)		Ziyaev et al. (1999)
				Nakano and Uchiyama (1956b)

Table 3.1 (cont'd)

Name of species	Plant parts[a]	Compounds	Type	References
	L & B	Liriodenine (17)		Ziyaev et al. (1999)
	L	Armepavine (29)		
		N-Norarmepavine (31)		
		Nornuciferine (28)		
		Magnocurarine (36)		
		Roemerine (22)		Ziyaev et al. (1995, 1999)
		Oxolaurenine (15)		Fukushima et al. (1996)
		Anonaine (7)		
		Lanuginosine (14)		
		Tyramine (4)		
	FR	Peonidin 3-rutinoside (49)	Flavonoid	Matsutani and Shiba (1975)
				Santamour (1966a)
	P	Rurin (58)		Hayashi and Ouchi (1948)
				Hayashi and Ouchi (1949a,b)
	B	Epieudesmin (69)	Lignan	DNP (1999)
	S	Yangambin (62)		Hirose et al. (1968)
	L			Iida et al. (1982a)
				Kamikado et al. (1975)
	ST	Sesamin (73)		Fukushima et al. (1996)
	L	Syringaresinol (61)		Iida et al. (1982a)
		Eudesmin (60)		
		Phillygenin (71)		
		Kobusinol A (89)		
		Spinescin (Kobusin) (74)		
	B	Kobusinol B (90)		DNP (1999)
		Eudesmin (60)		Brieskorn and Huber (1976),
		Syringin (185)		Iida et al. (1982a)
		Coniferin (184)		Kamikado et al. (1975)
				DNP (1999)
			Phenylpropanoid	Miyauchi and Ozawa (1998)
	ST	Kobusimin A (234)	Sesquiterpene	Fukushima et al. (1996)
	L	Kobusimin B (234)		Fukushima et al. (1996)
		9-Oxonerolidol (227)		Iida et al. (1982a)
		(S)-(E)-3-Hydroxy-1,6,10-phytatrien-9-one (228)	Others	
		Napthalene (254)		Azuma et al. (1996)

Species	Part	Compound	Type	Reference
Magnolia lennei DC.	F	Cyanidin 3-glucoside (47) Peonidin 3,5-diglucoside (50) Peonidin 3-rutinoside (49) Nictoflorin (54) Quercetin 3-glucoside (57) Quercetin (56) Rutin (58)	Flavonoid	Francis and Harborne (1966)
Magnolia liliiflora Desr.	L & B	Tyramine (4)	Alkaloid	Matsutani and Shiba (1975)
	B	Taspine (5)		Talpatra et al. (1982)
	L & B	(−)-Maglifloenone (162)	Neolignan	Talpatra et al. (1982)
	L	Burchellin (142) Denudatin A (144) Denudatin B (145) Furoenone (153) Liliflone (160) Piperinone (161) Liliflol A (158) Liliflol B (159)		Iida and Ito (1983), Talpatra et al. (1982) Iida and Ito (1983)
	L	Veraguensin (79) 9-Oxonerolidol (227) Liliflodione (255) Naphthalene (254) (E)-1,2,3,15-Tetranor-4,6,10-farnesatriene (247) 4,8,12-Trimethyl-1,3,7,11-tridecatetraene (248)	Lignan Sesquiterpene Others	Talpatra et al. (1982) DNP (1999) Iida and Ito (1983) Azuma et al. (1996) DNP (1999)
Magnolia liliifolia L	B	Magnocurarine (36) Salicifoline (3)	Alkaloid	Nakano (1953)
	L, B, & R	Cyanidin 3-glucoside (47) Peonidin 3-rutinoside (49) Peonidin 3,5-diglucoside (50) Nictoflorin (54) Quercetin (56) Quercetin 3-glucoside (57)		
	F	Rutin (58)	Flavonoid	Francis and Harborne (1966)

Table 3.1 (cont'd)

Name of species	Plant parts[a]	Compounds	Type	References
Magnolia macrophylla Desr.	L	Cyanidin (46)	Flavonoid	Santamour (1965b)
	B, L, FR & S	Rutin (58)		Plouvier (1943)
Magnolia mutabilis Regel	B	Magnolioside (42)	Coumarin	Plouvier (1968)
	B	Lanuginosine (14)	Alkaloid	Talpatra et al. (1975)
		Liriodenine (17)		
		Sesamin (73)	Lignan	
Magnolia obovata Thunb.	B	Magnocurarine (36)	Alkaloid	Sazaki (1921), Tomita et al. (1951)
	L	Liriodenine (17)		Ogiu and Morita (1953)
	L & R			Furmanowa and Jozefowicz (1980)
				Ito and Asai (1974)
				Furmanowa and Jozefowicz (1980)
	W	Anonaine (7)		Sashida et al. (1976)
	L			Ziyaev et al. (1999)
	L & R			Ito and Asai (1974)
	L	N-Acetylanonaine (8)		Ziyaev et al. (1999)
	W	Glaucine (19)		Hufford (1976)
	L & R			Ito and Asai (1974)
	L	Asimilobine (23)		Ziyaev et al. (1999)
				Ito and Asai (1974), Furmanowa and Jozefowicz (1980)
		Armepavine (29)		
		Magnoflorine (37)		
		N-Norarmepavine (31)		
		Nornuciferine (28)		
		Obovanine (12)		
		Reticuline (33)		
		Isolaurenine (20)		
		Roemerine (22)		
	B	Tyramine (4)		Ziyaev et al. (1999)
	FR	Peonidin 3,5-diglucoside (50)	Flavonoid	Matsutani and Shiba (1975)
	L	Rutin (58)		Santamour (1966a)
				Plouvier (1943)

	B	Neolignan	Honokiol (108)	Sugi (1930), Fujita et al. (1973)
			Magnolol (112)	Fukuyama et al. (1989, 1992)
			Clovanemagnolol (131)	
			Caryolanemagnolol (130)	Fukuyama et al. (1990b)
			Eudeshonokiol A (134)	
			Eudeshonokiol B (135)	
			Eudesobovatol A (138)	
			Eudesobovatol B (139)	
			Obovatol (105)	
	L		Obovatal (104)	Kwon et al. (1997), Ito et al. (1982)
			Obovaaldehyde (103)	
			Eudesmagnolol (133)	
			Magnolianin (106)	
	B	Phenylpropanoid	4-Hydroxy-3-methoxycinnamic acid (169)	Fukuyama et al. (1989, 1990a)
				Fukuyama et al. (1993)
	L		Magnoloside A (179)	Ziyaev et al. (1999)
	B		Magnoloside B (180)	Hasegawa et al. (1988)
			Magnoloside C (181)	DNP (1999)
		Sesquiterpene Monoterpene	Caryophyllene (226)	Fukuyama et al. (1992)
			Bornyl acetate (189)	
			Camphene (203)	
			α- and β-Pinene (195, 196)	
			β-Eudesmol (218)	
			Syringaldehyde (244)	
			Vanillin (245)	
		Others		Mori et al. (1997)
				Shinoda et al. (1976)
Magnolia officinalis Rehder & E.H. Wilson	L	Alkaloid	10-O-Demethylcryptaustoline (35)	Moriyasu (1996)
	B	Neolignan	Honokiol (108)	Chou et al. (1996), Fujita et al. (1973)
			Magnolol (112)	Sugi (1930), Fujita et al. (1973)
				Chou et al. (1996)
			Bornylmagnolol (129)	Konoshima et al. (1991)
			Dipiperitylmagnolol (132)	Yahara et al. (1991)
			Magnaldehyde B (116)	
			Magnaldehyde C (113)	
			Magnolignan A (119)	
			Magnolignan B (120)	
			Magnolignan C (121)	

Table 3.1 (cont'd)

Name of species	Plant parts[a]	Compounds	Type	References
		Magnolignan D (122)		Konoshima et al. (1991)
		Magnolignan E (123)		
		Magnaldehyde D (114)		
		Magnotriol A (127)		
		Magnotriol B (128)		
		Magnaldehyde E (115)		
		Magnolignan F (124)		
		Magnolignan G (107)		
		Magnolignan H (126)		
		Magnolignan I (125)		
		Piperitylhonokiol (136)		
		Piperitylmagnolol (137)		
		Randainal (118)		
		1-(4-hydroxy-3-methoxyphenyl)-2-[4-(ω-hydroxypropyl)-2-methoxyphenoxy]propane-1,3-diol (100)		
		3,5'-Diallyl-2'-hydroxy-4-methoxybiphenyl (110)		
		Syringaresinol (61)	Lignan	Yahara et al. (1991)
		Syringaresinol glucoside (63)		
		Eugenol methyl ether (177)	Phenylpropanoid	
		Sinapic aldehyde (175)		
		β-Eudesmol (218)	Sesquiterpene	Baek et al. (1992) Konoshima et al. (1991)
Magnolia parviflora Siebold & Zucc.	B	Magnoflorine (37) Magnocurarine (36)	Alkaloid	Nakano and Uchiyama (1956a)
Magnolia pterocarpa Roxb.	L	Imperatorin (45) Eudesmin (60) Fargesin (70) Sesamin (73)	Coumarin Lignan	Talpatra et al. (1983)
Magnolia rostrata W.W. Sm.	B	Honokiol (108) Magnolol (112) β-Eudesmol (218)	Neolignan Sesquiterpene	Yan (1979)

Magnolia saliicifolia Maxim.	B	Alkaloid	Saliciofine (3)	Tomita and Nakano (1952a)

Species	Part	Class	Compounds	References
Magnolia saliicifolia Maxim.	B	Alkaloid	Saliciofine (3)	Tomita and Nakano (1952a)
	L		Armepavine (29), Asimilobine (23), Liriodenine (17), Magnoflorine (37), Magnocurarine (36), Nornuciferine (28), N-Norarmepavine (31), Roemerine (22)	Furmanowa and Jozefowicz (1980)
	F & FR		Coclaurine (32)	Watanabe *et al.* (1981), Boudet (1973)
	FR	Flavonoid	Reticuline (33)	Santamour (1966a)
	FB		Peonidin 3-rutinoside (49), Astragalin (52), Nictoflorin (54), Tiliroside (55), Magnosalicin (82)	Tsuruga *et al.* (1991)
	L	Lignan	Magnosalin (99), Magnoshinin (98)	Tsuruga *et al.* (1991), Mori *et al.* (1987); Kikuchi *et al.* (1983)
	L & B	Phenylpropanoid	*trans*-Anethole (168)	Fujita and Fujita (1974)
	L		Eugenol methyl ether (177)	Azuma *et al.* (1996)
	L & F		Methylisoeugenol (172)	Kelm *et al.* (1997)
	FB		3,4-Methylenedioxycinnamyl alcohol (173), 1-(2,4,5-Trimethoxyphenyl)-1,2-propanedione (165)	Tsuruga *et al.* (1991)
	L & B		*cis*-Anethole (167), Eugenol (176), Isoeugenol (170), Safrole (178)	Fujita and Fujita (1974); Nagashima *et al.* (1981)
	FB	Sesquiterpene	Acetoside (183), Costunolide (204), Parthenolide (212)	Tsuruga *et al.* (1991)
	FR		Caryophyllene (226)	Kelm *et al.* (1997)
	L & B	Monoterpene	Camphene (203), Fenchone (198), Nerol (193), α-Terpenol (199)	Fujita and Fujita (1974)

Table 3.1 (cont'd)

Name of species	Plant parts[a]	Compounds	Type	References
		Terpinen-4-ol (200)		Nagashima et al. (1981)
		Citral A (190)		
		Citral B (191)		
		Cineol (194)		
		Camphor (195)		
		Geraniol (192)		
		Linalool (202)		
		Limonene (201)		
		α- and β-Pinene (196, 197)		
	L & B	Anisaldehyde (240)	Others	Fujita and Fujita (1974)
		p-Cymene (242)		
	FB	Asarylaldehyde (241)		Tsuruga et al. (1991)
		Vetraic acid (246)		
Magnolia sinensis (Rehder & E.H. Wilson) Stapf	F	Cyanidin 3-glucoside (47)	Flavonoids	Francis and Harborne (1966)
		Nictoflorin (54)		
		Peonidin (48)		
		Peonidin 3-rutinoside (49)		
		Quercetin (56)		
		Quercetin 3-glucoside (57)		
		Rutin (58)		
Magnolia sieboldii K. Koch	L	Magnoporphine (25)	Alkaloid	DNP (1999)
	FR	Cyanidin 3-glucoside (47)	Flavonoids	Santamour (1966a)
	F	Nictoflorin (54)		Francis and Harborne (1966)
		Peonidin (48)		
		Peonidin 3-rutinoside (49)		
		Peonidin 3,5-diglucoside (50)		
		Quercetin (56)		
		Quercetin 3-glucoside (57)		
		Rutin (58)		
	ST & B	Syringin (185)	Phenylpropanoid	Park et al. (1996)
		Syringinoside (186)	Sesquiterpene	Park et al. (1996, 1997)
		Costunolide (204)		

Species	Plant part	Compound	Type	Reference
	R & ST	15-Acetoxycostunolide (205)		Tada et al. (1982)
Magnolia soulangeana Soul.-Bod	B	Oxolaureline (15)	Alkaloid	Ziyaev et al. (1975)
	L & B	Anonaine (7), Liriodenine (17), Roemerine (22)		Ziyaev et al. (1999)
	FR	Peonidin 3,5-diglucoside (50)	Flavonoid	Santamour (1966a)
	F	Rutin (58)		Plouvier (1943)
	FB	Aurein (140), Cyclohexadienone (143), Denudatin A (144), Denudatin B (145), Saulangianin I (164)	Neolignan	Abdallah (1993)
		Eudesmin (60), Veraguensin (79)	Lignan	
	L	Magnolin (64), Soulangianolide A (206), Soulangianolide B (216)	Sesquiterpene	El-Feraly (1983)
		Chromanols (250), α-Tocopherol (249)	Others	Lichtenthaler (1965)
Magnolia sprengeri Pamp.	L	Magnosprengerine (2)	Alkaloid	DNP (1999)
Magnolia stellata (Siebold & Zucc.) Max.	B	Salicifoline (3)	Alkaloid	Tomita and Nakano (1952c)
	L	Cyanidin (46), Quercetin (56)	Flavonoid	Santamour (1965b)
	P	Magnostellin B (163)	Neolignan	Weevers (1930)
	B	Magnostellin A (81), Eudesmin (60), Piperitol (72), Sesamin (73), Spinescin (Kobusin) (74), Vomifoliol (253)	Lignan	Iida et al. (1983)
			Others	Brieskorn and Huber (1976)
Magnolia thompsoniana de Vos	ST, L, FR & S	Rutin (58)	Flavonoid	Plouvier (1943)
Magnolia tomentosa Roxb.	L	Napthalene (254)	Others	Azuma et al. (1996)

Table 3.1 (cont'd)

Name of species	Plant parts[a]	Compounds	Type	References
Magnolia tripetala L.	L	Armepavine (29) Asimilobine (23) Liriodenine (17) Magnocurarine (36) Magnoflorine (37) N-Norarmepavine (31) Nornuciferine (28) Roemerine (22)	Alkaloid	Furmanowa and Jozefowicz (1980)
	L	Cyanidin (46)	Flavonoid	Santamour (1965b)
	FR			Santamour (1966a)
	F	Isoquercetin (51) Rutin (58)		Santamour (1966b)
	B	Honokiol (108) Magnolol (112)	Neolignan	Fujita et al. (1973)
Magnolia virginiana L.	FR	Peonidin-3-rutinoside (49)	Flavonoid	Santamour (1966a)
	F	Honokiol (108)	Neolignan	Chandra and Nair (1995)
	B	Magnolol (112)		Fujita et al. (1973)
	F	4,4′-Diallyl-2,3′-dihydroxybiphenylether (101) 3,5′-Diallyl-2′-hydroxy-4-methoxybiphenyl (110)		Chandra and Nair (1995)
	L	Costunolide (204) Costunolact-12β-ol (208) Constunolact-12β-ol acetal dimer (207) Parthenolide (212) Trifloculoside (232)	Sesquiterpene	Nitao et al. (1991) Song et al. (1998)

Species	Part	Compounds	Class	Reference
Magnolia watsonii Hook.	L	Asimilobine (23) Liriodenine (17) Honokiol (108) Magnolol (112) Obovatol (105) Watsonol A (229) Watsonol B (230)	Alkaloid Neolignan Sesquiterpene	Ito *et al.* (1984a,b)
Magnolia yulan Desf.	P	Quercetin (56)	Flavonoid	Weevers (1930)
	L & F	Rutin (58)		Plouvier (1943)
Magnolia species	P	N,N-Dimethyllindcarpine (26) Norglaucine (27) N-Methyllindcarpine (21) Norushinsunine (11)	Alkaloid	DNP (1999)
	W	Syringin (185)	Phenylpropanoid	Plouvier (1962)
	P	β-Carotene (237) α-Carotene-5,6-epoxide (236) Lutein (238) Lutein 5,6-epoxide (239)	Tetraterpene	Demuth and Santamour (1978)
		Quinic acid (251) Shikimic acid (253)	Others	Boudet (1973)

[a] B = barks; D = duramen; F = flowers; FB = flower-buds; FR = fruits; L = leaves; P = petals; R = roots; RB = root-barks; S = seeds; ST = stems; W = wood.

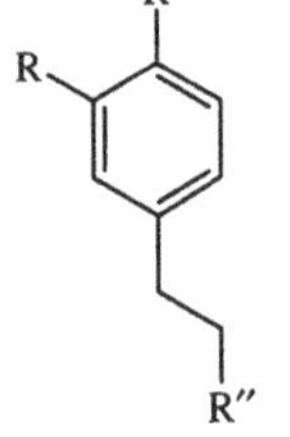

	R	R′	R″
Candicine (**1**)	OH	H	$N(Me)_3^+$
Magnosprengerine (**2**)	OMe	OH	$N(Me)_2$
Salicifoline (**3**)	OH	OMe	$N(Me)_3^+$
Tyramine (**4**)	H	OH	NH_2

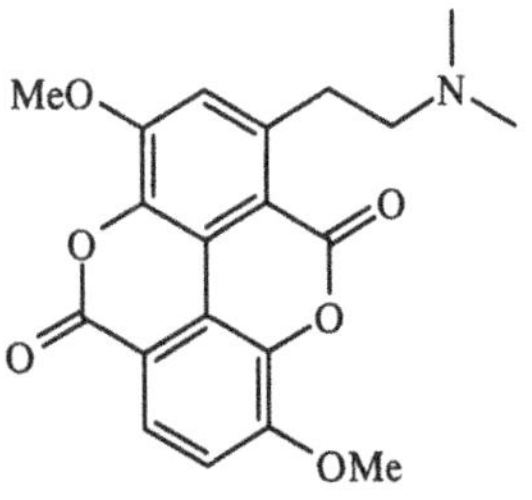

Taspine (**5**) 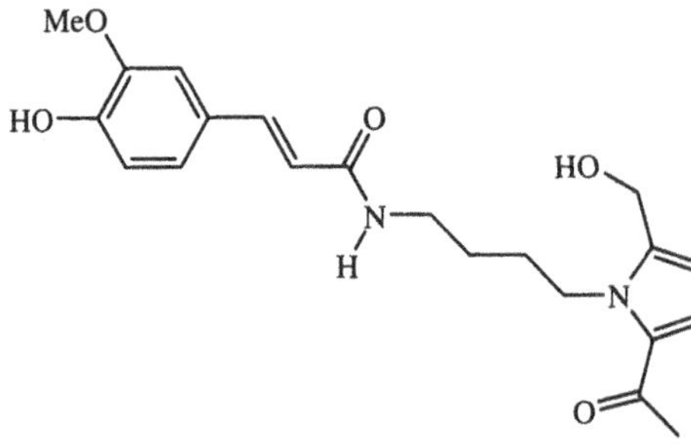   Magnolamide (**6**)

Figure 3.1 Amino alkaloids from *Magnolia*.

alkaloids, and so is the genus *Magnolia*. *Magnolia* produces mainly isoquinoline-type alkaloids, the majority of which are aporphine/noraporphine derivatives. In total 40 different alkaloids (Figures 3.1 to 3.4), which can be further classified under amino (1–6), aporphine/noraporphine (7–28), benzylisoquinoline (29–37) and bisbenzyliso-quinoline alkaloids (38–40), have been reported from 23 different species of *Magnolia* (Table 3.1). *M. obovata* appears to be the richest source and produces 15 different alkaloids (Table 3.1). From the phytochemical data available to date, it is evident that not all species of this genus are capable of producing alkaloids. Liriodenine (17), which is also known as oxoushinsunine, micheline B or spermatheridine, is the most common alkaloid in *Magnolia*, and has been found in at least 14 different species. Different classes of *Magnolia* alkaloids are described under respective subheadings.

3.2.1.1 Amino alkaloids

Unlike other alkaloids where the nitrogen atom is a member of a heterocyclic ring, the nitrogen atom of an amino alkaloid is located in an amino group outside the ring (Samuelsson, 1999). In fact, most of them are the building blocks of larger alkaloids. Among the amino alkaloids (1–6) found in *Magnolia*, salicifoline (3) is the most common one. Salicifoline was first isolated from *M. salicifolia* almost half a century ago (Tomita and Nakano, 1952a) and later found in six other species of *Magnolia* (Table 3.1).

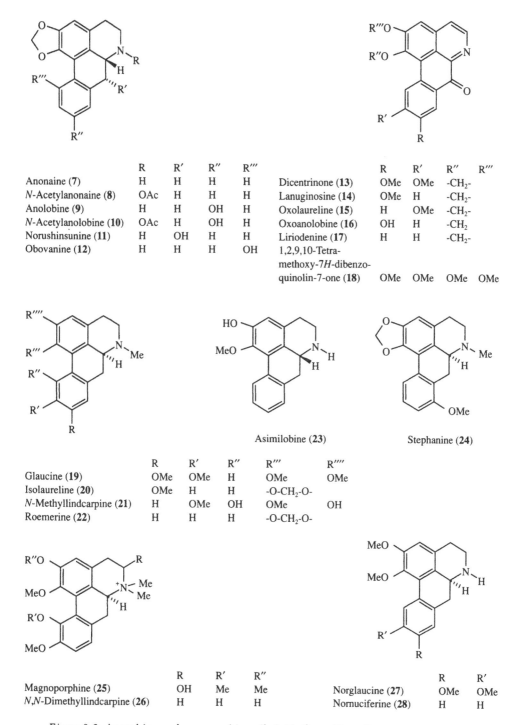

	R	R'	R''	R'''
Anonaine (**7**)	H	H	H	H
N-Acetylanonaine (**8**)	OAc	H	H	H
Anolobine (**9**)	H	H	OH	H
N-Acetylanolobine (**10**)	OAc	H	OH	H
Norushinsunine (**11**)	H	OH	H	H
Obovanine (**12**)	H	H	H	OH

	R	R'	R''	R'''
Dicentrinone (**13**)	OMe	OMe	-CH₂-	
Lanuginosine (**14**)	OMe	H	-CH₂-	
Oxolaureline (**15**)	H	OMe	-CH₂-	
Oxoanolobine (**16**)	OH	H	-CH₂	
Liriodenine (**17**)	H	H	-CH₂-	
1,2,9,10-Tetra-methoxy-7*H*-dibenzo-quinolin-7-one (**18**)	OMe	OMe	OMe	OMe

Asimilobine (**23**)

Stephanine (**24**)

	R	R'	R''	R'''	R''''
Glaucine (**19**)	OMe	OMe	H	OMe	OMe
Isolaureline (**20**)	OMe	H	H	-O-CH₂-O-	
N-Methyllindcarpine (**21**)	H	OMe	OH	OMe	OH
Roemerine (**22**)	H	H	H	-O-CH₂-O-	

	R	R'	R''
Magnoporphine (**25**)	OH	Me	Me
N,N-Dimethyllindcarpine (**26**)	H	H	H

	R	R'
Norglaucine (**27**)	OMe	OMe
Nornuciferine (**28**)	H	H

Figure 3.2 Aporphine and noraporphine alkaloids from *Magnolia*.

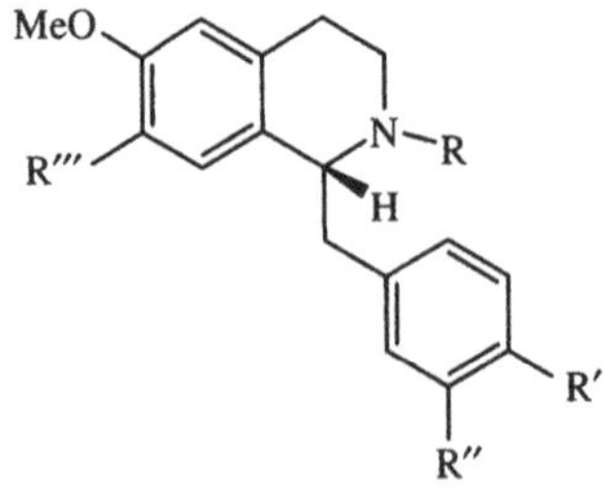

	R	R'	R''	R'''
Aremepavine (**29**)	Me	OH	H	OMe
1-*O*-Methylarmepavine (**30**)	Me	OMe	H	OMe
N-Norarmepavine (**31**)	H	OH	H	OMe
Coclaurine (**32**)	H	OH	H	OH
Reticuline (**33**)	Me	OMe	OH	OH

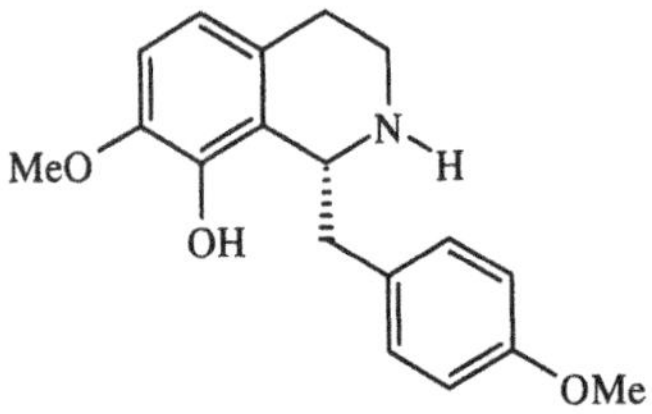

Magnococline (**34**)

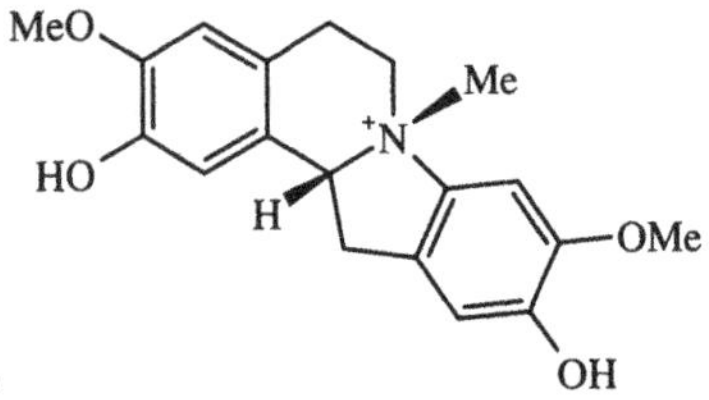

10-*O*-Demethylcryptaustoline (**35**)

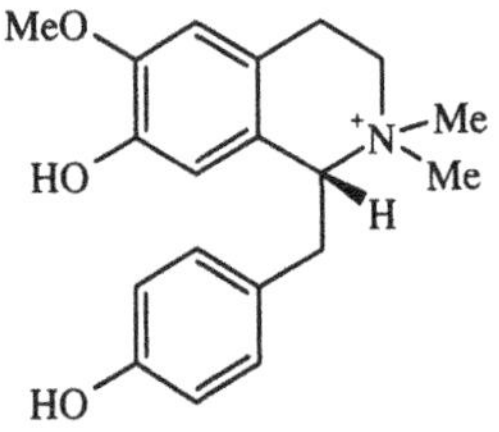

Magnocurarine (**36**)

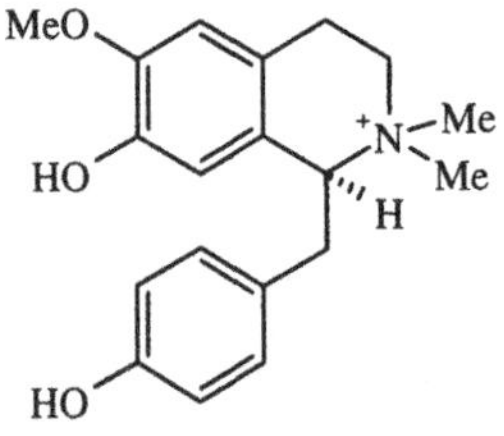

Magnoflorine (**37**)

Figure 3.3 Benzylisoquinoline alkaloids from *Magnolia*.

3.2.1.2   Isoquinoline alkaloids

Isoquinoline alkaloids (7–40) form the major group of alkaloids obtained from *Magnolia*. These alkaloids are biosynthetically derived from the amino acid tyrosine. They can be subclassified into aporphine- and noraporphine-type (Figure 3.2), benzylisoquinoline-type (Figure 3.3) and bisbenzylisoquinoline-type (Figure 3.4).

3.2.1.2.1   APORPHINE AND NORAPORPHINE ALKALOIDS

The aporphine and noraporphine alkaloids represent a large group of isoquinoline alkaloids, and are restricted to the families Annonaceae, Berberidaceae, Lauraceae,

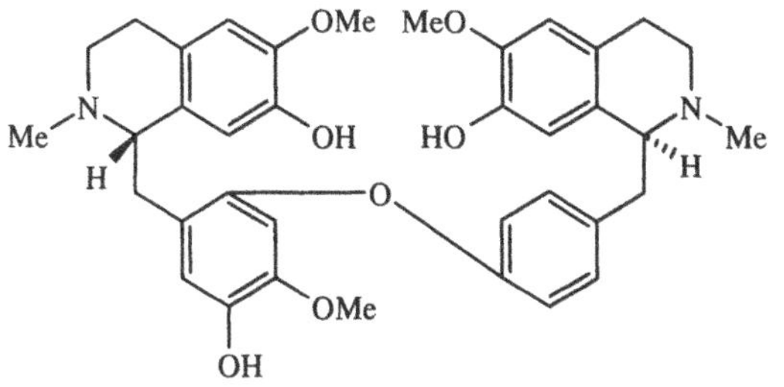

Magnolamine (**38**)

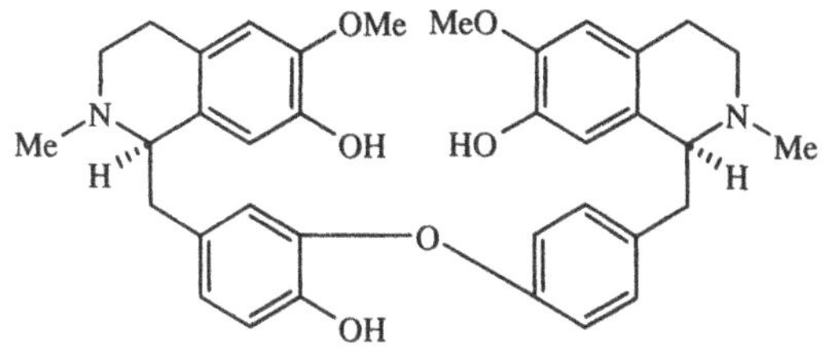

Magnoline (**39**)

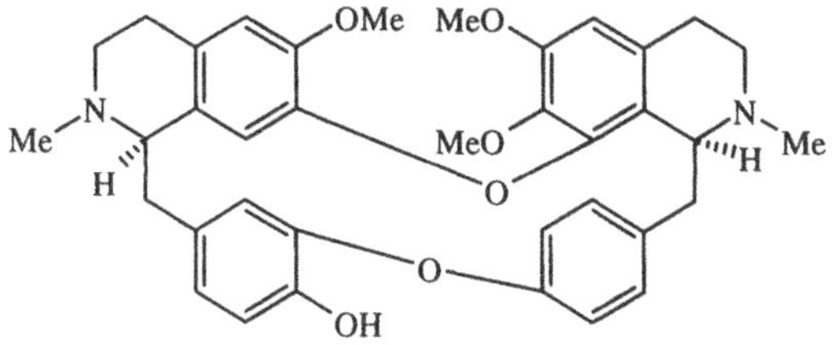

Oxycanthine (**40**)

Figure 3.4 Bisbenzylisoquinoline alkaloids from *Magnolia*.

Magnoliaceae, Menispermaceae, Monimiaceae, Ranunculaceae, Papaveraceae and Rhamnaceae (Guinaudeau *et al.*, 1975). Several aporphine/noraporphine-type alkaloids (7–28) have been isolated from various species of *Magnolia* (Table 3.1). Anonaine (**7**), asimilobine (**23**), liriodenine (**17**), nornuciferine (**28**), and roemerine (**22**) are the five most widely distributed aporphine/noraporphine alkaloids in this genus.

3.2.1.2.2 BENZYLISOQUINOLINE ALKALOIDS

In total, nine *Magnolia* alkaloids (29–37) fall into this category. The most common two—magnocurarine (**36**) and magnoflorine (**37**), first isolated, respectively, from *M. obovata* (Sazaki, 1921) and *M. grandiflora* (Nakano, 1954a,b), also occur in nine other species. While armepavine (**29**), *N*-norarmepavine (**31**) and reticuline (**33**) have been found in more than one species of *Magnolia*, methylarmepavine (**30**), coclaurine (**32**), magnococline (**34**) and 10-demethylcryptaustoline (**35**) have been reported from only one species so far (Table 3.1).

3.2.1.2.3 BISBENZYLISOQUINOLINE ALKALOIDS

The bisbenzylisoquinoline alkaloids are dimers of benzylisoquinolines connected by one to three ether linkages formed by phenol coupling. Over 270 members of this class of alkaloids have been found in various plants (Samuelsson, 1999), but from *Magnolia*, only three such alkaloids (38–40) have been reported to date. Both magnoline (39) and magnolamine (38), which are derived from two units of benzylisoquinoline alkaloids through a single biphenyl ether linkage formation, were isolated from *M. fuscata* (Proskurnina and Orekhov, 1938, 1940; Komissarova, 1945). Two biphenyl ether linkages are present in oxycanthine (40), which was found only in *M. compressa* (DNP, 1999). While biphenyl ether formation is not very common among *Magnolia* alkaloids, it is a common feature in *Magnolia* neolignans (see Figure 3.10). This suggests that these biphenyl ethers are perhaps not artefacts originating during the extraction and isolation process; rather, *Magnolia* plants possibly produce the necessary enzymes responsible for this type of biphenyl ether formation. In a recent study, it has been shown that the enzyme berbamunine synthase in *Berberis stolonifera* catalyses the reaction leading to the formation of berbamunine, the optical isomer of which is also known as magnoline (39) (Samuelsson, 1999).

3.2.2   Coumarins

Coumarins and furanocoumarins are derivatives of 5,6-benzo-2-pyrone. Production of coumarins seems not to be favoured in *Magnolia*, and only four species have so far been reported to have coumarins. Scoparone (41), magnolioside (42) and imperatorin (45) were isolated, respectively, from *M. compressa*, *M. macrophylla* and *M. pterocarpa* (Table 3.1). *M. grandiflora* produced two coumarins: 6-methoxy-7-hydroxycoumarin (43) and 6,8-dimethoxy-7-hydroxycoumarin (44). All five isolated coumarins (Figure 3.5), except

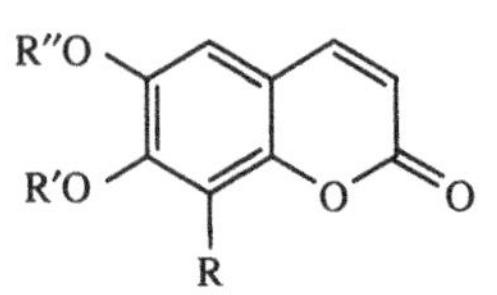

	R	R′	R″
Scoparone (41)	H	Me	Me
Magnolioside (42)	H	Me	Glucosyl
6-Methoxy-7-hydroxycoumarin (43)	H	H	Me
6,8-Dimethoxy-7-hydroxycoumarin (44)	OMe	H	OMe

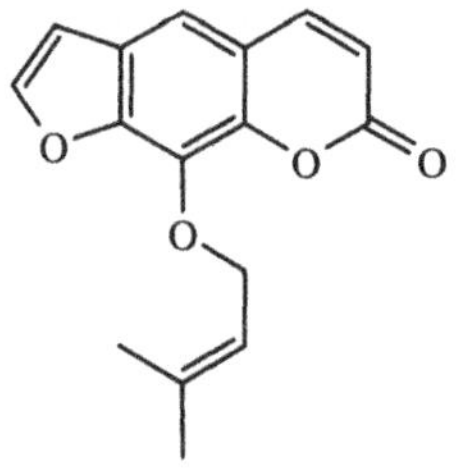

Imperatorin (45)

Figure 3.5 Coumarins from *Magnolia*.

45, are simple coumarins. Imperatorin (45) is a furanocoumarin with an 8-prenyloxy group. Biosynthesis of coumarins begins with the oxidation of *trans*-cinnamic acid to *o*-coumaric acid, and is followed by glucosylation and isomerisation leading to the formation of *cis-o*-coumaric acid glucoside, which gives coumarin by ring closure. Further hydroxylation, methoxylation, glucosylation and prenylation on the coumarin nucleous produce compounds 1–4. Hydroxylation at C-7 and prenylation (from isopentyl diphosphate) at C-6 of the coumarin molecule, followed by cyclisation, leads to the formation of a psoralen skeleton; finally, incorporation of a prenyloxy group at C-8 completes the structure of the furanocoumarin (45) found in *Magnolia*.

3.2.3 Flavonoids

Flavonoids, biosynthesised via a combination of the shikimic acid and acylpolymalonate pathways, form the largest group of naturally occurring phenols, which structurally are derivatives of 1,3-diphenylpropane. All flavonoids are derived from phenylalanine, the starting material for flavonoid biosynthesis. Two of the most common classes of flavonoids are anthocyanidins and flavones. All flavonoids (46–58) isolated from at least 21 different *Magnolia* species fall into these two classes. Structurally they are cyanidin, peonidin, kaempherol and quercetin derivatives (Figure 3.6). Glycosylation, especially at C-3, appears to be the most common feature among the *Magnolia* flavonoids. Rutin (58) and peonidin rutinoside (49) are the two most widely distributed flavonoids in *Magnolia* and have been reported from more than 10 species. The only isoflavonoid, isoquercetin (51) has been found in *M. tripetala* and *M. fraseri* (Santamour, 1966b). No other plant secondary metabolites, only flavonoids, have been reported from *M. ashei*, *M. cordata*, *M. faseri*, *M. lennii*, *M. sinensis*, *M. thomsoniana* and *M. yulan* (Table 3.1). However, this does not necessarily imply that these species are not capable of producing other types of compounds; in fact, it is possible that further thorough phytochemical investigation will result in the isolation of other types of compounds from these species.

3.2.4 Lignans

Lignans are optically active dimeric natural products formed essentially by the union of two phenylpropane units, biosynthesised through the shikimic acid pathway and widely distributed in the plant kingdom. Magnoliaceae is one of the 70 plant families that are renowned for producing lignans with varying levels of abundance. Lignans form one of the major groups of secondary metabolites found in the genus *Magnolia*. To date, 41 different lignans, mainly belonging to the classes tetrahydrofurofuran, tetrahydrofuran and aryltetralin, have been reported from 16 different species of this genus (Table 3.1, Figures 3.7 to 3.9). Eudesmin (60) and sesamin (73), both of tetrahydrofurofuran-type, occur widely in *Magnolia* and have been isolated from nine different species of *Magnolia*. The two top lignan-producing species are *M. compressa* and *M. fargesii* from which, respectively, 13 and 14 different types of lignans have been reported (Table 3.1).

3.2.4.1 Tetrahydrofurofuran-type lignans

Tetrahydrofurofuran-type lignans (Figure 3.7) are formed by two C_6–C_3 units β-β′ linked, and two additional oxygen bridges (Massanet *et al.*, 1988). This is one of the

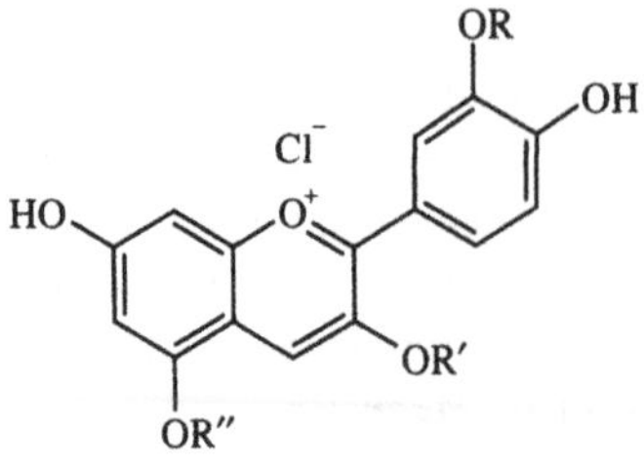

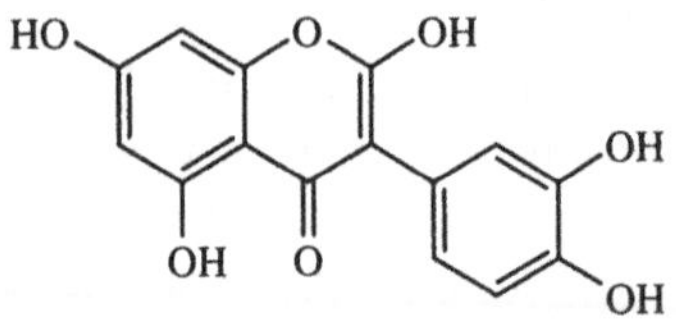

	R	R′	R″
Cyanidin (**46**)	H	H	H
Cyanidin 3-glucoside (**47**)	H	Glucosyl	H
Peonidin (**48**)	Me	H	H
Peonidin 3-rutinoside (**49**)	Me	Rutinosyl	H
Peonidin 3,5-diglucoside (**50**)	Me	Glucosyl	Glucosyl

Isoquercetin (**51**)

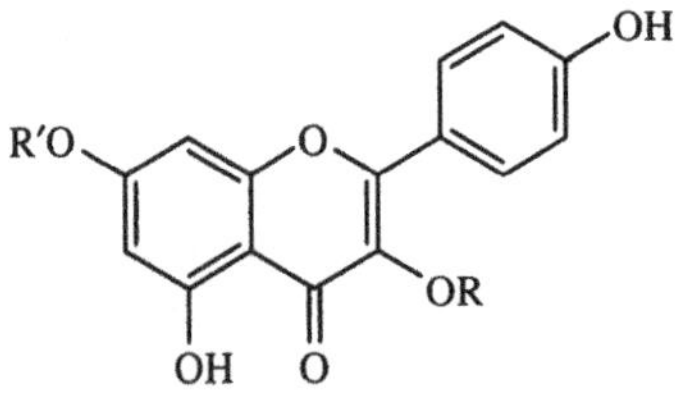

	R	R′
Astragalin (**52**)	Glucosyl	H
Buddlenoid A (**53**)	H	(6-*O*-Coumaroyl)-glycosyl
Nictoflorin (**54**)	Rutinosyl	H
Tiliroside (**55**)	(6-*O*-Coumaroyl)-glycosyl	H

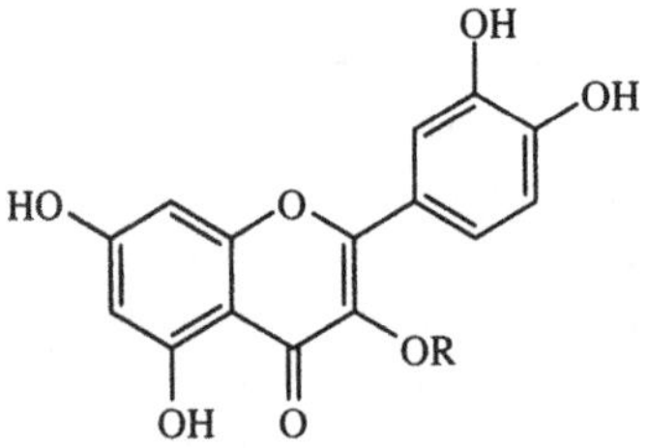

	R
Quercetin (**56**)	H
Quercetin 3-glucoside (**57**)	Glucosyl
Rutin (**58**)	Rutinosyl

Figure 3.6 Flavonoids from *Magnolia*.

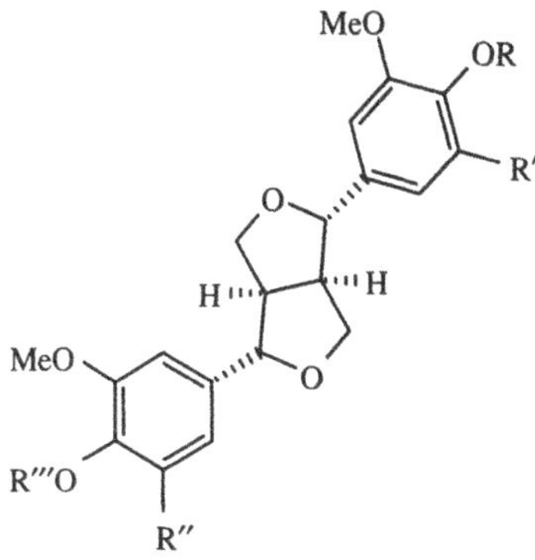

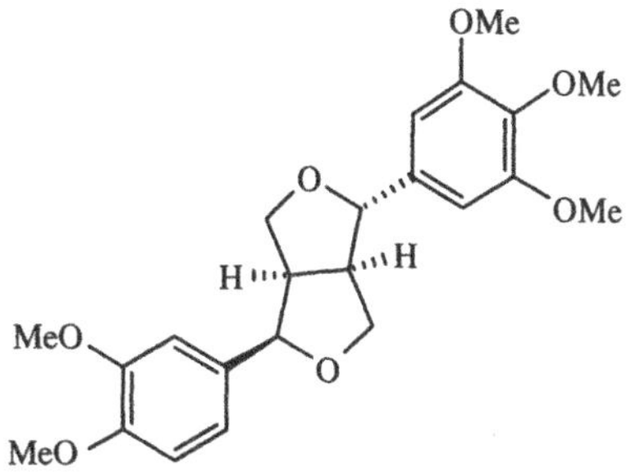

	R	R'	R''	R'''
Pinoresinol (**59**)	H	H	H	H
Eudesmin (**60**)	Me	H	H	Me
Syringaresinol (**61**)	H	OMe	OMe	H
Yangambin (**62**)	Me	OMe	OMe	Me
Syringaresinol glucoside (**63**)	Glucosyl	OMe	OMe	Me
Magnolin (**64**)	Me	OMe	H	Me
3-*O*-Demethylmagnolin (**65**)	H	OMe	H	Me

Epimagnolin A (**66**)

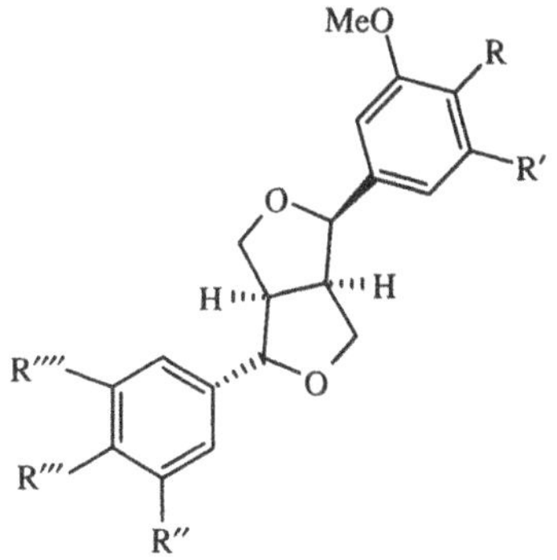

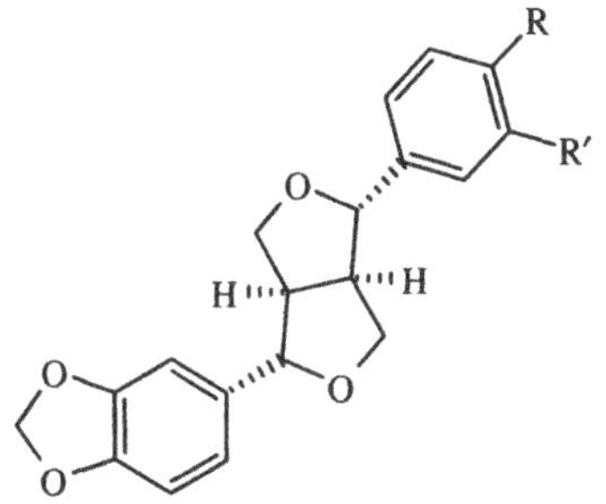

	R	R'	R''	R'''	R''''		R	R'
Lirioresinol A (**67**)	OH	OMe	OMe	OH	OMe	Piperitol (**72**)	OH	OMe
Lirioresinol A dimethylether (**68**)	OMe	OMe	OMe	OMe	OMe	Sesamin (**73**)	-O-CH₂-O-	
Epieudesmin (**69**)	OMe	H	H	OMe	OMe	Spinescin (**74**)	OMe	OMe
Fargesin (**70**)	OMe	H	H	-O-CH₂-O				
Phillygenin (**71**)	OMe	H	H	OH	OMe			

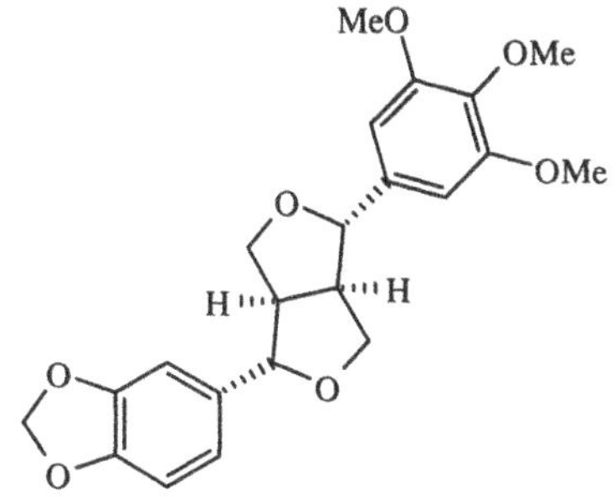

Aschantin (**75**)

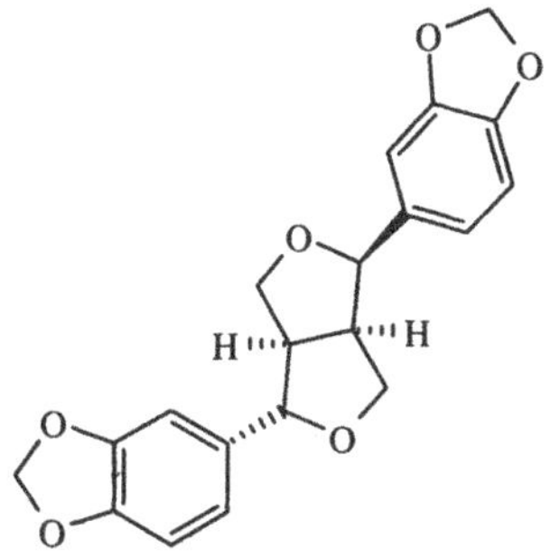

Episesamin (**76**)

Figure 3.7 Tetrahydrofurofuran-type lignans from *Magnolia*.

two major types of lignans found in *Magnolia*. Eighteen different lignans (59–76), which mainly differ from each other by the presence or absence of hydroxyl or methoxyl groups at different positions of the two phenyl rings and also in their stereochemistry, have been isolated from this genus. Among these lignans, eudesmin (60), fargesin (70), pinoresinol (59), sesamin (73), spinescin (also known as kobusin) (74) and syringaresinol (61) occur in more than four different species.

3.2.4.2   *Tetrahydrofuran-type lignans*

Tetrahydrofuran-type lignans (Figure 3.8) are formed by two C_6–C_3 units β-β' linked, and an additional oxygen bridge (Massanet *et al.*, 1988). A total of 19 different tetrahydrofuran-type lignans (77–95) have been reported from *Magnolia* (Figure 3.8, Table 3.1). Veraguensin (79) is the most common one and has been isolated from *M. acuminata* (Doskotch and Flom, 1972), *M. denudata* (Iida *et al.*, 1982b), *M. liliflora* (Talpatra *et al.*, 1982) and *M. soulangeana* (Abdallah, 1993).

3.2.4.3   *Aryltetralin-type and other lignans*

Aryltetralins (Figure 3.9) are formed by two C_6–C_3 units β-β' linked, and an additional C-2, C-α' bond generating a tetralin (Massanet *et al.*, 1988). Guaiacin (96), magnoliadiol (97) and magnoshinin (98), isolated respectively from *M. kachirachirai* (Ito *et al.*, 1984a,b), *M. fargesii* (Miyazawa *et al.*, 1995, 1996) and *M. salicifolia* (Kikuchi *et al.*, 1983), are of aryltetralin-type. Magnosalin (99), which does not fall into any of the three types of lignans described here, was isolated from the leaves of *M. salicifolia* (Kikuchi *et al.*, 1983).

3.2.5   *Neolignans*

Neolignans are derived from two C_6–C_3 units linked otherwise than β-β' (Massanet *et al.*, 1988). Neolignan is the largest group of secondary metabolites found in *Magnolia*. Since the discovery of the pharmacologically active neolignan magnolol (112) (Sugi, 1930; Sarker, 1997) from the stem bark of *M. officinalis* and *M. obovata*, significant attention has been given to isolating, identifying and assessing the biological activity of this type of secondary metabolite from the genus *Magnolia*. As a result, 65 different types of neolignans have been reported from 17 different species of *Magnolia* so far. All these compounds are now described under subgroups: biphenyl ether, biphenyl, terpenyl and other neolignans (Figures 3.10 to 3.13). *M. officinalis* and *M. obovata* are the two richest sources of such compounds, and *M. officinalis* alone produces 23 different neolignans.

3.2.5.1   *Biphenyl ether-type neolignans*

This type of neolignan is formed from two units of phenylpropane through phenol coupling. There are eight neolignans of this type reported from *Magnolia* (Figure 3.10). In compounds 101, 102, 104 and 105, an ether linkage is formed between two phenyl groups of two phenylpropane units. Compound 100 does not strictly fall into this group as the ether linkage is formed between the phenyl ring of one phenylpropane unit and the side chain carbon of the other. In the case of obovaldehyde (103), biphenyl ether formation is observed between a phenylpropane unit and a benzaldehyde

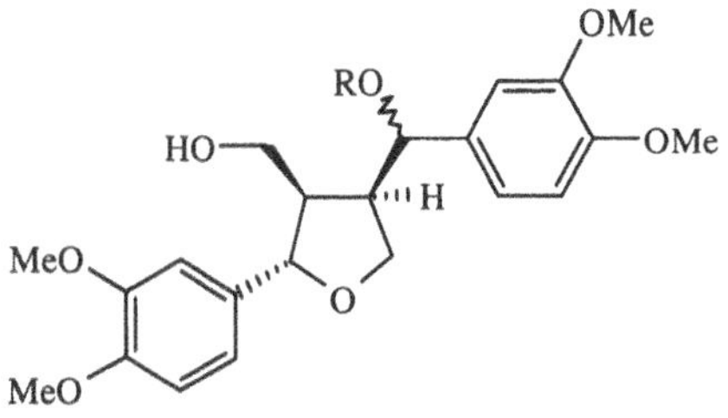

Biondinin B (**77**) R = H
Biondinin E (**78**) R = Me

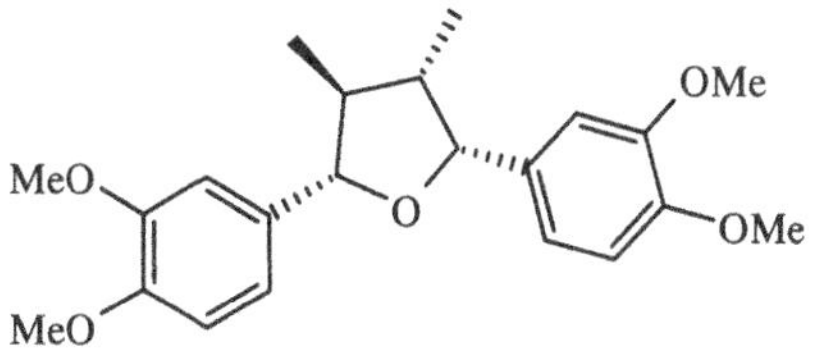

Veraguensin (**79**)

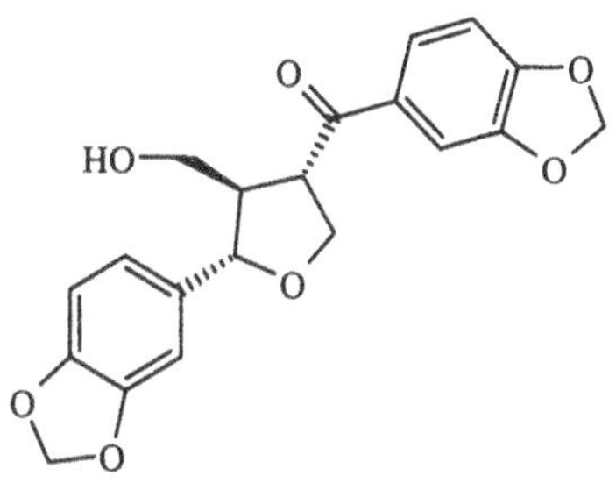

Sesaminone (**80**)

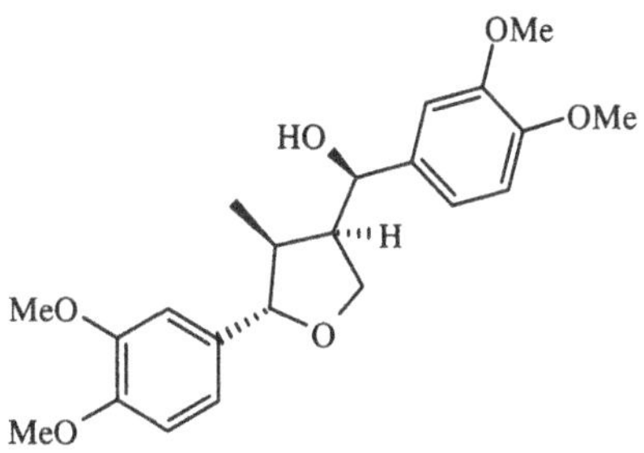

Magnostellin A (**81**)

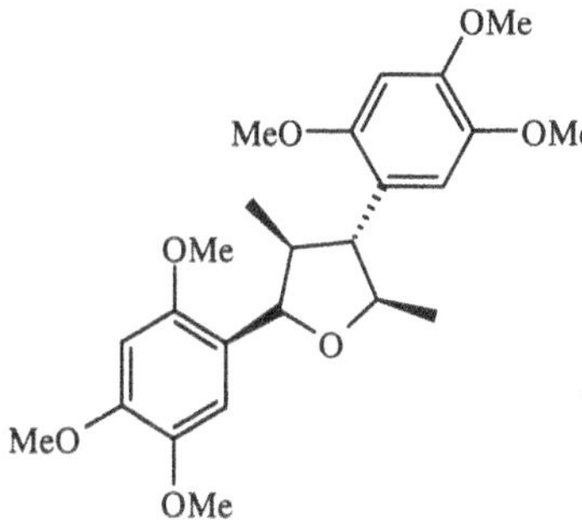

Magnosalicin (**82**)

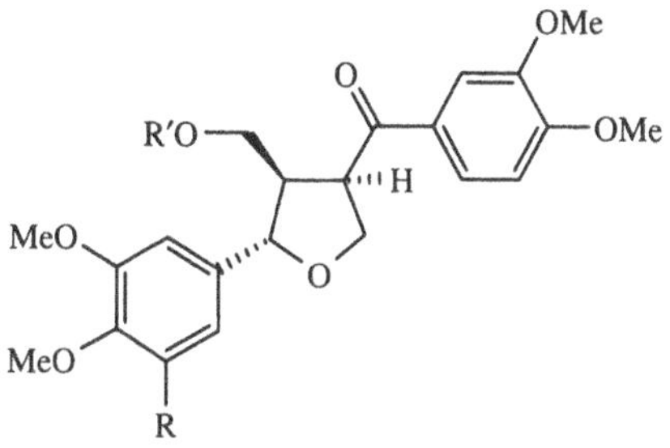

	R	R'
Magnone A (**83**)	H	H
Magnone B (**84**)	OMe	Me

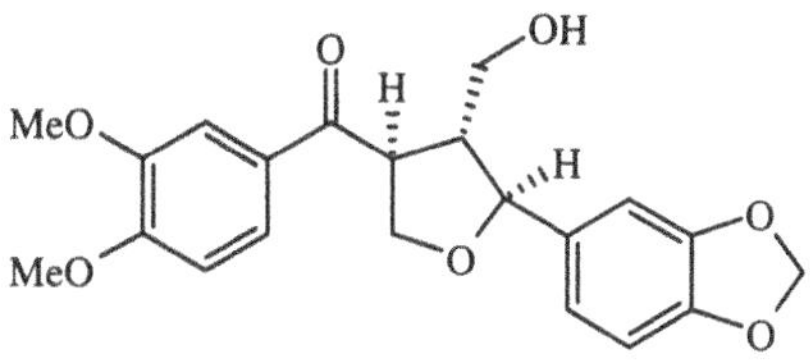

Magnolone (**85**)

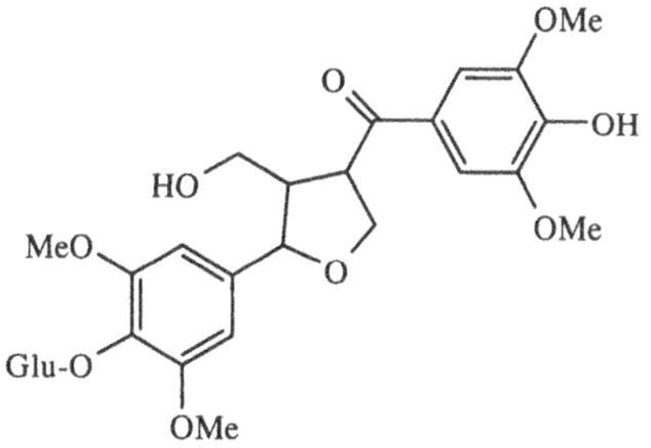

Magnolenin C (**86**)

Figure 3.8 Tetrahydrofuran lignans from *Magnolia*.

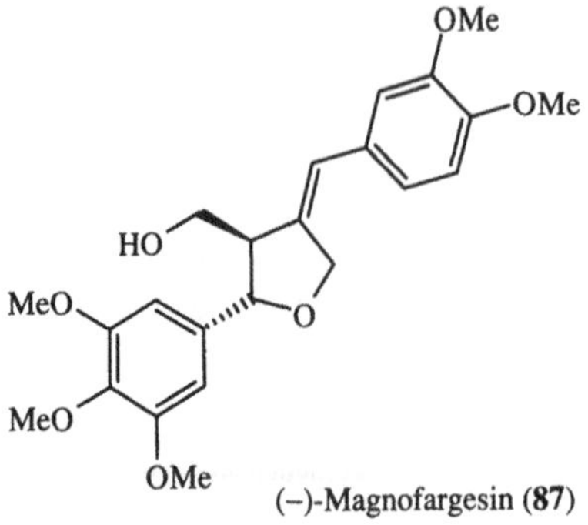

(−)-Magnofargesin (**87**)

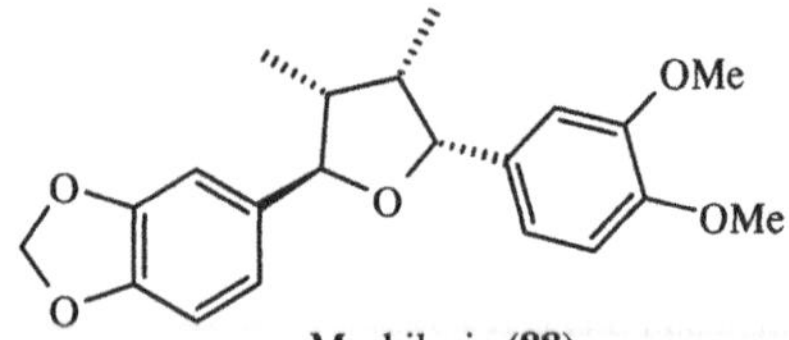

Machilusin (**88**)

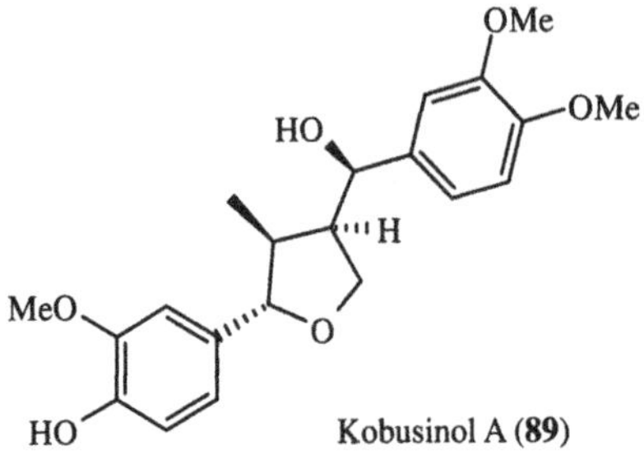

Kobusinol A (**89**)

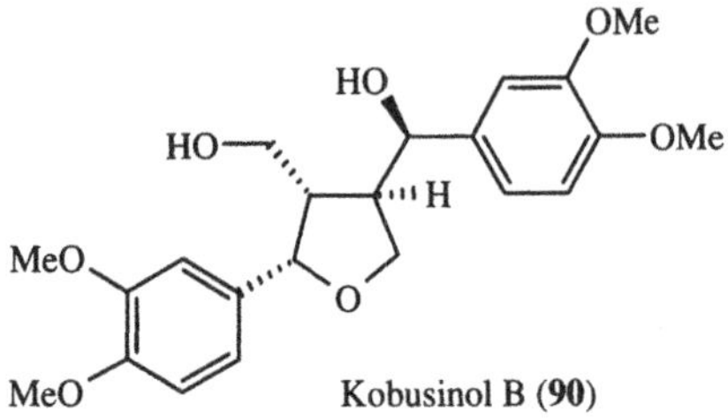

Kobusinol B (**90**)

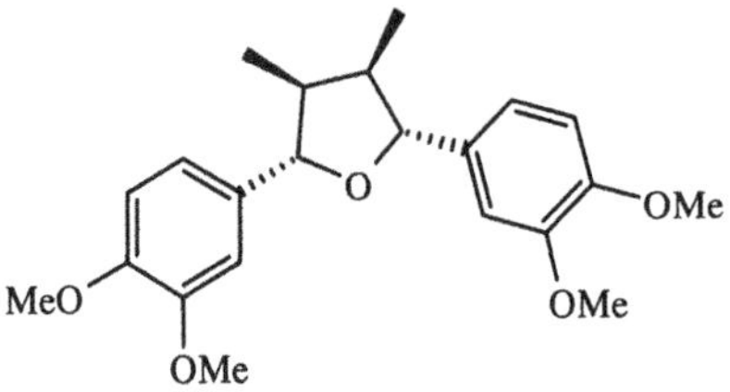

Galgravin (**91**)

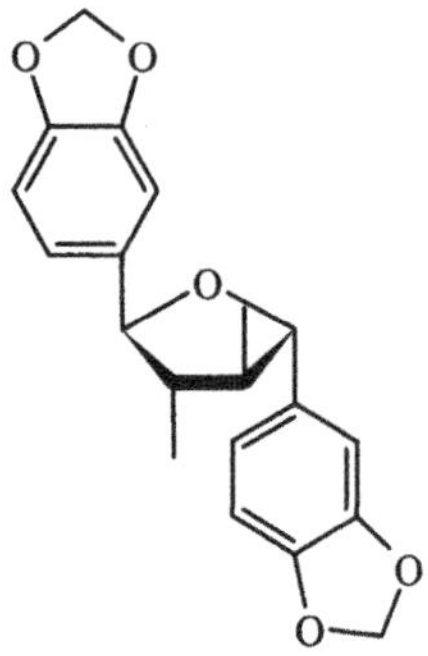

(+/−)-Galbacin (**92**)

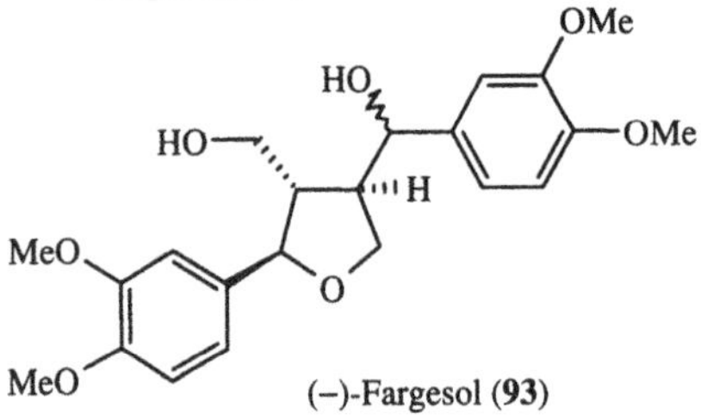

(−)-Fargesol (**93**)

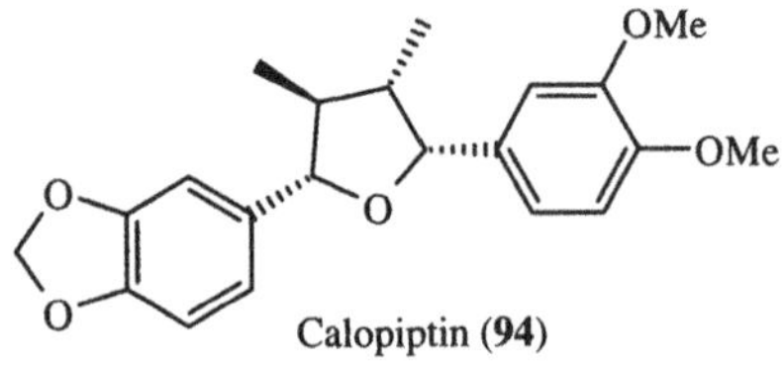

Calopiptin (**94**)

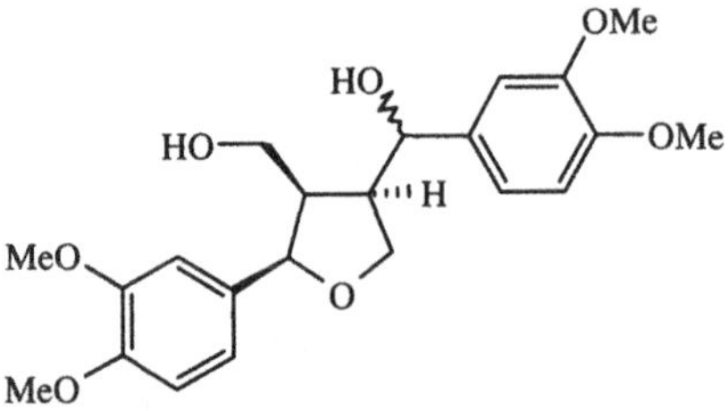

(7*R*,7′*m*,8*R*,8′*R*)-7′,9-Dihydroxy-3,3′,4,4′-tetramethoxy-7,9′-epoxylignan (**95**)

Figure 3.8 (*continued*)

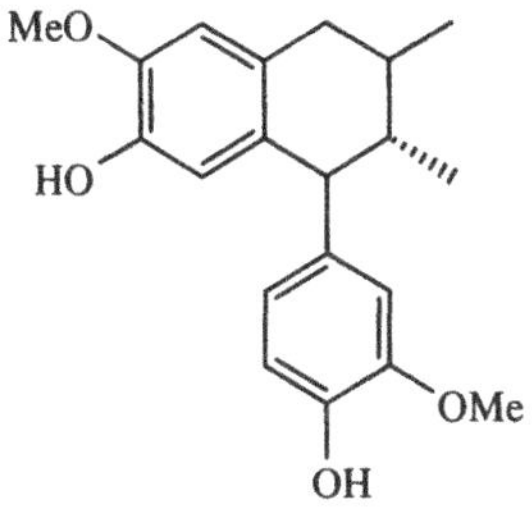

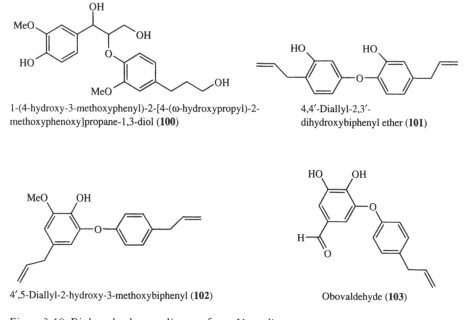

Guaiacin (**96**)

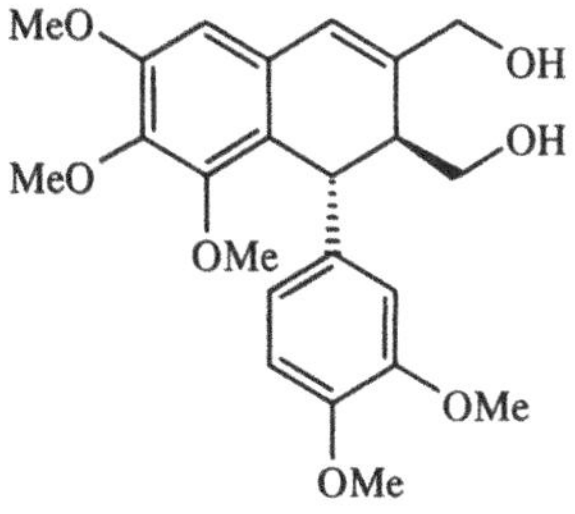

(+)-Magnoliadiol (**97**)

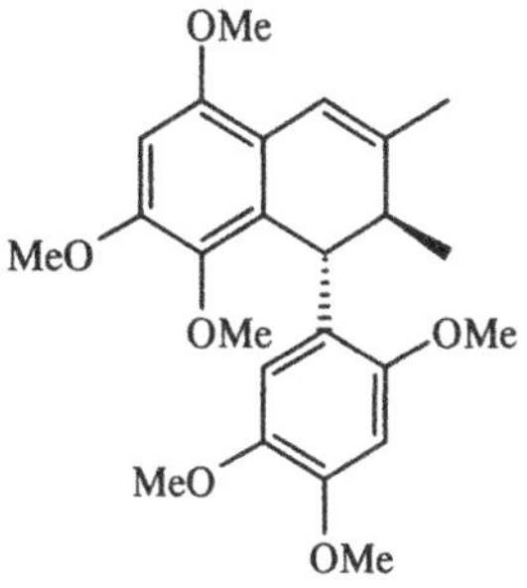

Magnoshinin (**98**)

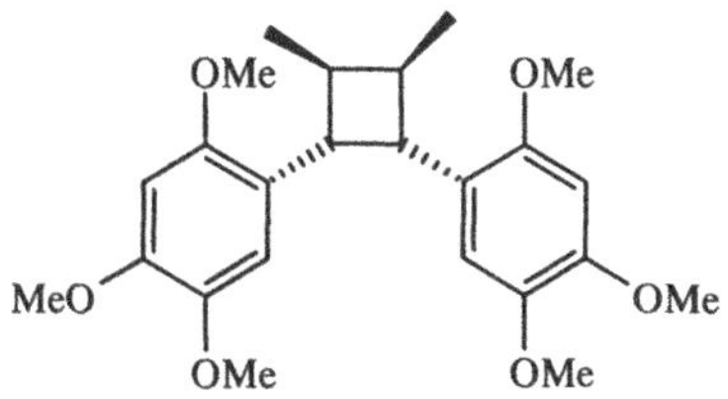

Magnosalin (**99**)

Figure 3.9 Aryltetralins and other lignans from *Magnolia*.

1-(4-hydroxy-3-methoxyphenyl)-2-[4-(ω-hydroxypropyl)-2-methoxyphenoxy]propane-1,3-diol (**100**)

4,4′-Diallyl-2,3′-dihydroxybiphenyl ether (**101**)

4′,5-Diallyl-2-hydroxy-3-methoxybiphenyl (**102**)

Obovaldehyde (**103**)

Figure 3.10 Biphenyl ether neolignans from *Magnolia*.

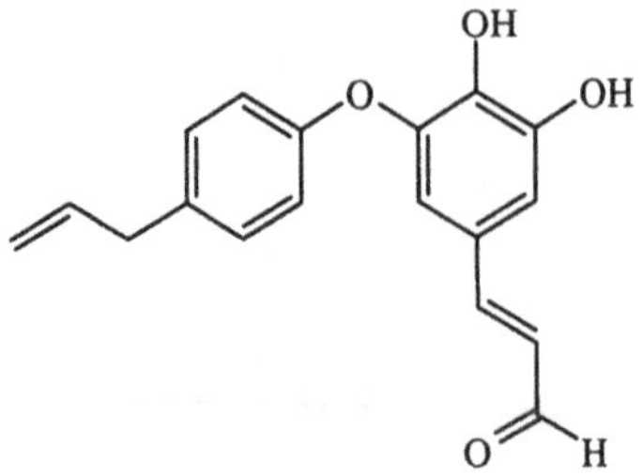

Obovatal (**104**)

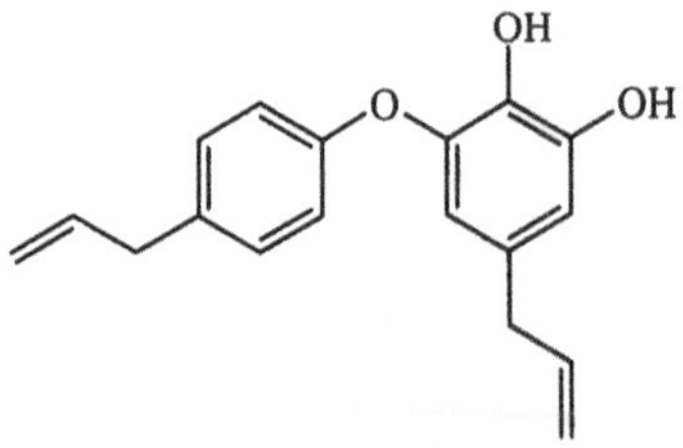

Obovatol (**105**)

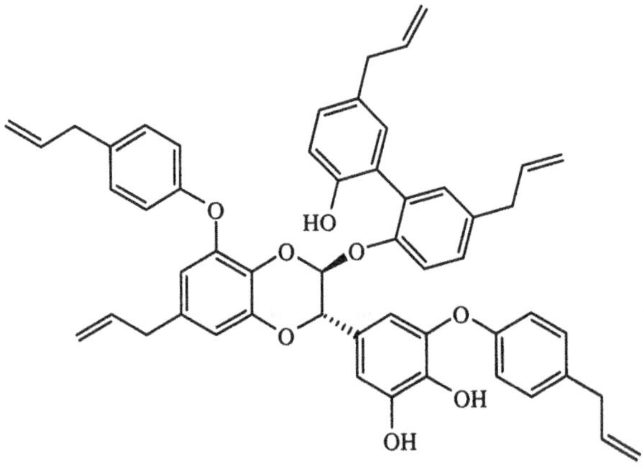

Magnolianin (**106**)

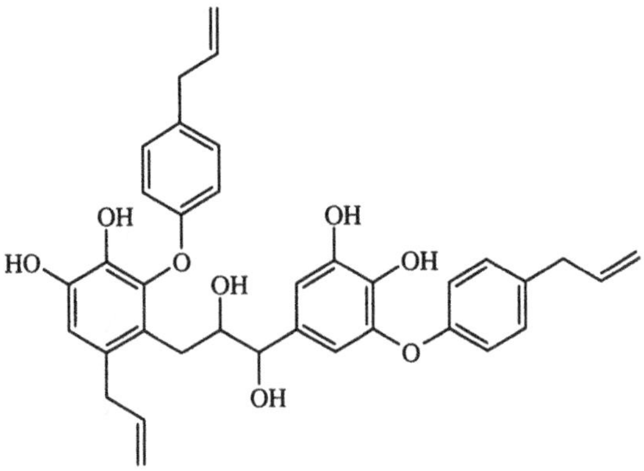

Magnolignan G (**107**)

Figure 3.10 (continued)

derivative. Magnolianin (106), isolated from the bark of *M. obovata* (Fukuyama *et al.*, 1993), is a trimeric neolignan formed from two units of biphenyl ether neolignans and another unit of magnolol (112), which is a biphenyl neolignan. Similarly, magnolignan G (107) from *M. officinalis* (Yahara *et al.*, 1991) is a dimeric neolignan derived from coupling between two units of biphenyl ether neolignans.

3.2.5.2 *Biphenyl-type neolignans*

Honokiol (108) and magnolol (112), two positional isomers, are the two most commonly distributed biphenyl neolignans in *Magnolia*. They were first isolated from *M. officinalis* and *M. obovata*, but are also found in several other *Magnolia* species. Two units of phenylpropane couple through carbon–carbon bond formation between two phenyl groups to produce biphenyl neolignans. To date, 21 neolignans of this category (Figure 3.11), most of which are structurally similar to mognolol/honokiol and only differ from each other in the presence or absence of hydroxyl groups, methoxyl groups on the phenyl ring(s) or slight modification(s) in the side chain(s) (108–122), have been reported from *Magnolia*. Hydroxylation on the propenyl side chain is quite common and is evident from the presence of magnaldehyde C (113) and magnolignans A–D (119–122) in *Magnolia*. In magnolignan E (123), hydroxylations followed by cyclisation of one of the propenyl side chains form an additional five-membered ring fused with the phenyl ring. Magnolignans F and I [(124) and (125)] are dimeric biphenyl neolignans. In magnolignan H (126), dimerisation occurs between a biphenyl ether-type and a biphenyl-type neolignan. The biphenyl carbon–carbon bond formation in magnotriol A and B [(127) and (128)] is between a phenylpropanoid and a *p*-hydroxyphenol. Compounds (113–116) and (118–128) have been reported exclusively from the bark of *M. officinalis* (Yahara *et al.*, 1991).

3.2.5.3 *Terpenyl neolignans*

Terpenyl neolignans are formed from the coupling between a neolignan unit and a terpenoid unit. In *Magnolia*, this coupling is mainly formed through an ether linkage, but carbon–carbon bond formation is also observed (Figure 3.12). Honokiol (108), magnolol (112) and obovatol (105) are the only neolignans that take part in such coupling. The terpenoid unit(s) attached to these neolignans are either monoterpene, as in bornylmagnolol (129), dipiperitylmagnolol (132), piperitylhonokiol (136) and piperitylmagnolol (137), or sesquiterpene as in caryolanemagnolol (130), clovane-magnolol (131), eudesmagnolol (133), eudeshonokiol A (134), eudeshonokiol B (135), eudesobovatol A (138) and eudesobovatol B (139). Terpenyl neolignans do not seem to be widely distributed in *Magnolia*; their distribution, in fact, is very much restricted to *M. officinalis* and *M. obovata* (Table 3.1). A total of 11 terpenyl neolignans have been reported from these two species. It is interesting to note that all monoterpenyl neolignans (129, 132, 136, 137) have only been isolated from *M. officinalis*, whereas all sesquiterpenyl neolignans (130, 131, 133, 134, 135, 138, 139) are exclusively produced by *M. obovata*.

3.2.5.4 *Other neolignans*

In addition to the above three subclasses of neolignans, at least 25 other neolignans of diverse structural types have been isolated from *Magnolia* (Figure 3.13). Both denudatin

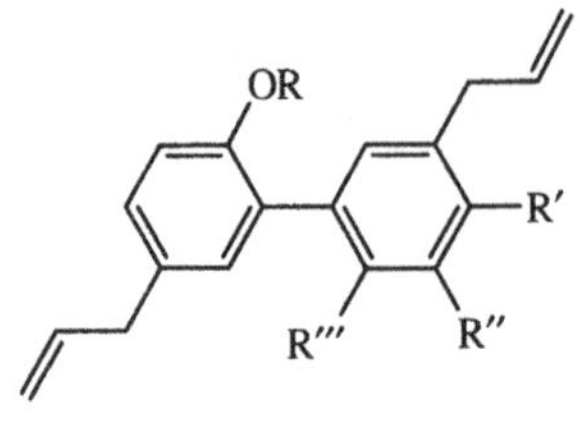

	R	R'	R''	R'''
Honokiol (**108**)	H	OH	H	H
4-*O*-methylhonokiol (**109**)	Me	OH	H	H
3,5'-Diallyl-2'-hydroxy-4-methoxybiphenyl (**110**)	H	OMe	H	H

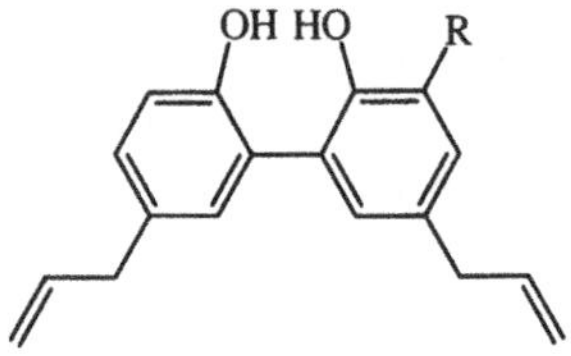

5,5'-Diallyl-2,2'-dihydroxy-3-methoxybiphenyl (**111**)   R = OMe
Magnolol (**112**)   R = H

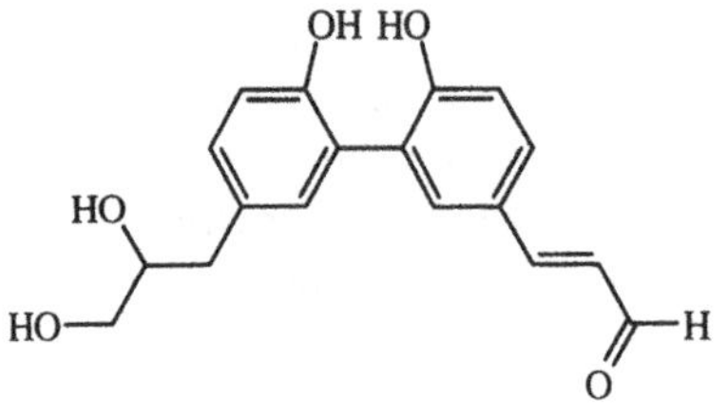

Magnaldehyde C (**113**)

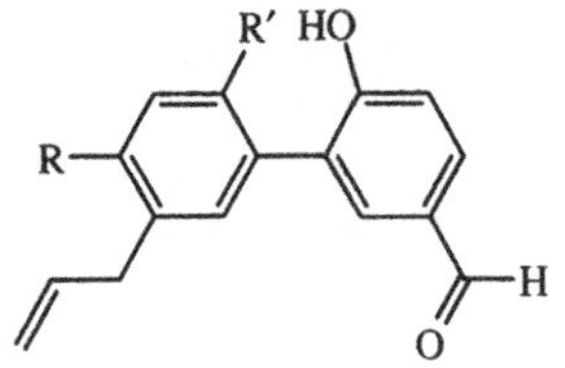

	R	R'
Magnaldehyde D (**114**)	H	OH
Magnaldehyde E (**115**)	OH	H

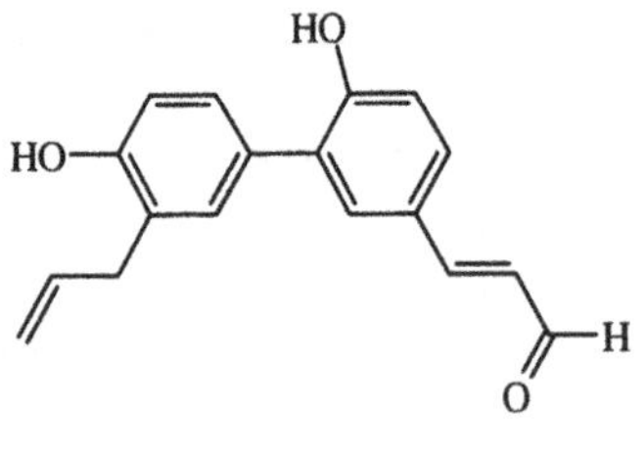

Magnaldehyde B (**116**) 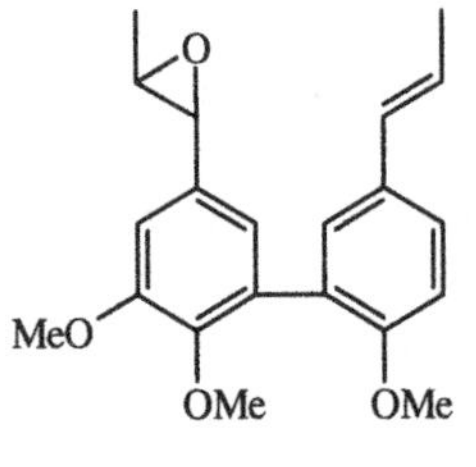   Acuminatin (**117**) 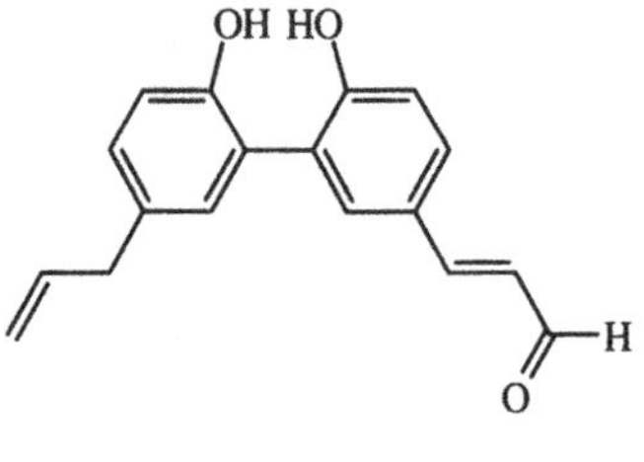   Randinal (**118**)

Figure 3.11 Biphenyl neolignans from *Magnolia*.

A and B (144 and 145), occur in *M. denudata* (Iida *et al.*, 1982b), *M. liliflora* (Iida and Ito, 1983; Talpatra *et al.*, 1982) and *M. soulangeana* (Abdallah, 1993). Denudatin B has also been isolated from the flower buds of *M. fargesii* (Yu *et al.*, 1990). While burchellin (142), futoenone (153) and licarin A (154) have been reported from at least two different species of *Magnolia*, all other neolignans are limited to a single occurrence (Table 3.1).

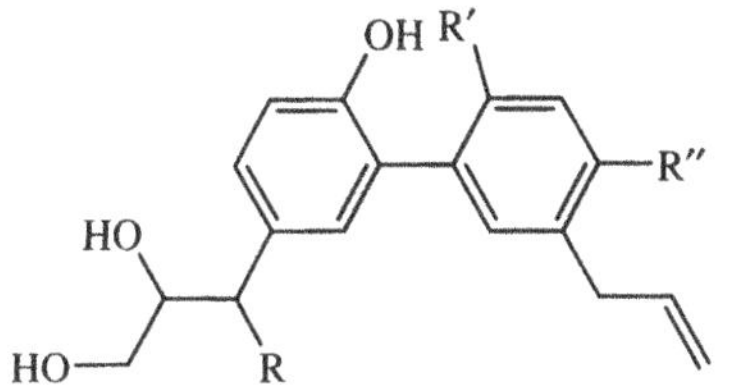

		R	R'	R''
Magnolignan A (**119**)		H	OH	H
Magnolignan B (**120**)		OH	OH	H
Magnolignan C (**121**)		H	H	OH
Magnolignan D (**122**)		OMe	H	OH

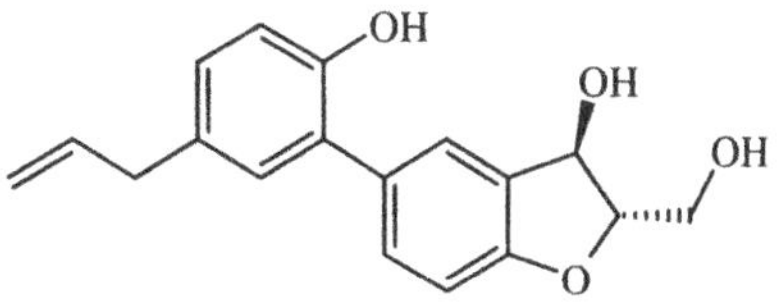

Magnolignan E (**123**)

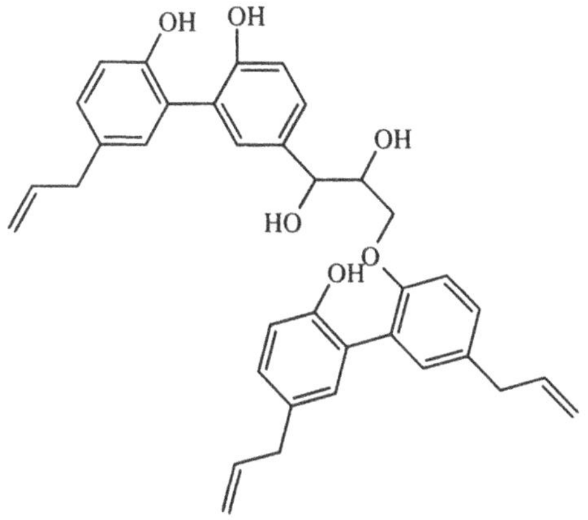

Magnolignan F (**124**)

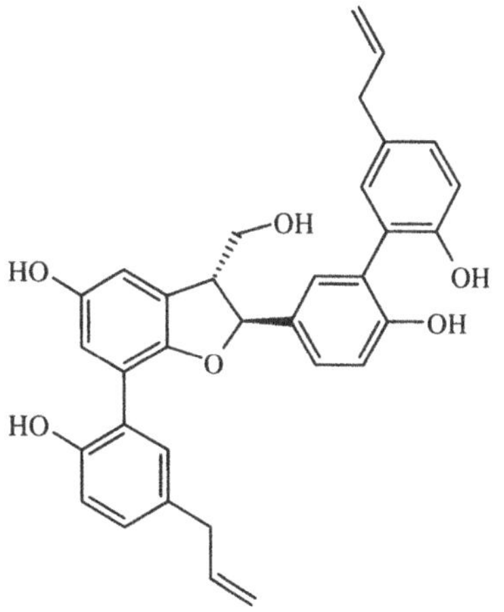

Magnolignan I (**125**)

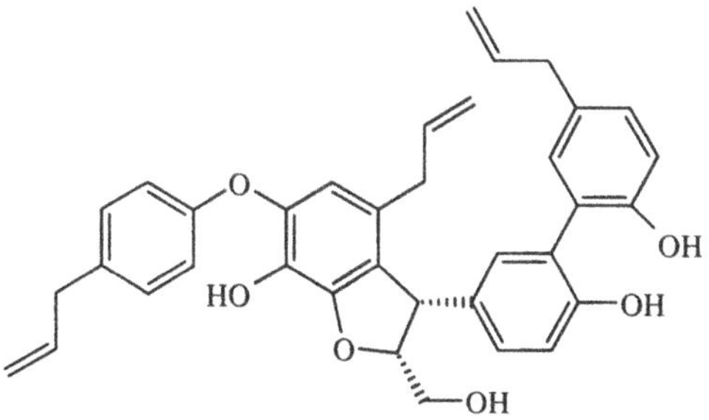

Magnolignan H (**126**)

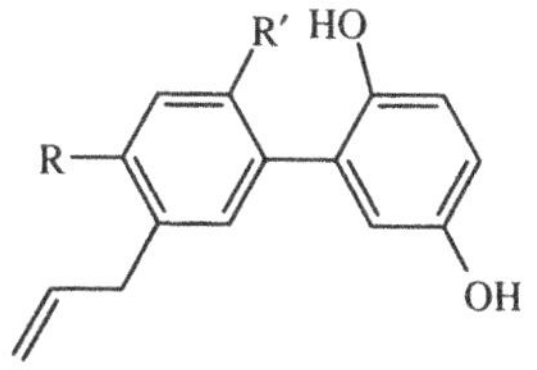

	R	R'
Magnotriol A (**127**)	H	OH
Magnotriol B (**128**)	OH	H

Figure 3.11 (continued)

3.2.6 *Phenylpropanoids*

Phenylpropanoids are aromatic compounds with a propyl side chain attached to the benzene ring. They are the building blocks or precursors of many plant secondary metabolites, e.g. coumarins, flavonoids, lignans, neolignans, etc. These compounds can be formed directly from phenylalanine and usually carry oxygenated substituent(s) on the aromatic ring. Cinnamic acid, a typical phenylpropane, is derived from phenylalanine through enzyme-catalysed (phenylalanine ammonia-lyase) deamination. As *Magnolia*

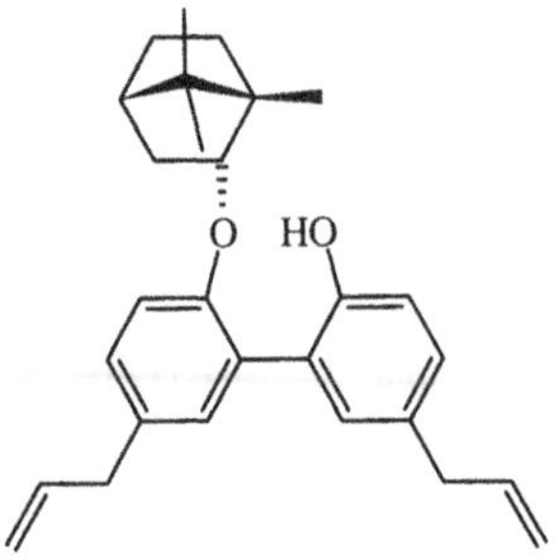

Bornylmagnolol (**129**) 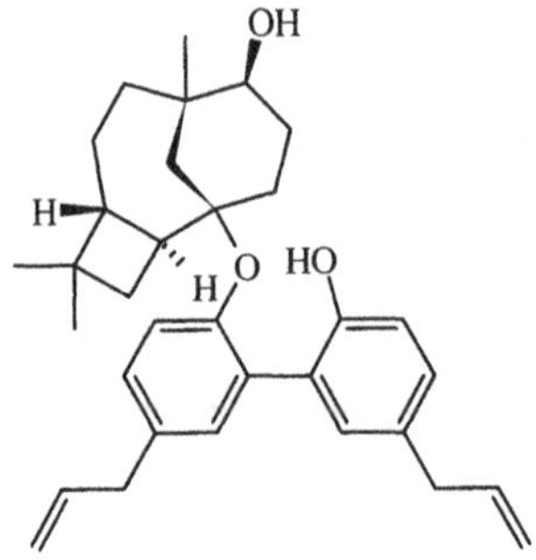   Caryolanemagnolol (**130**) 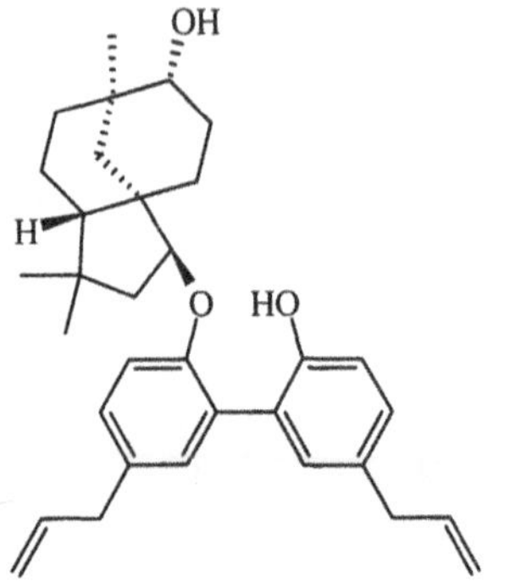   Clovanemagnolol (**131**)

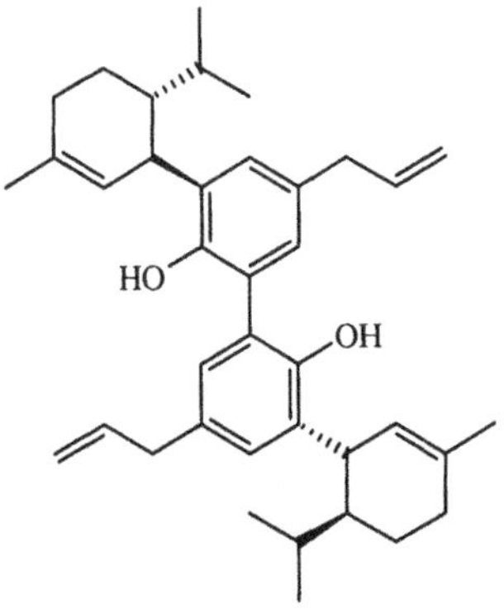

Dipeperitylmagnolol (**132**) 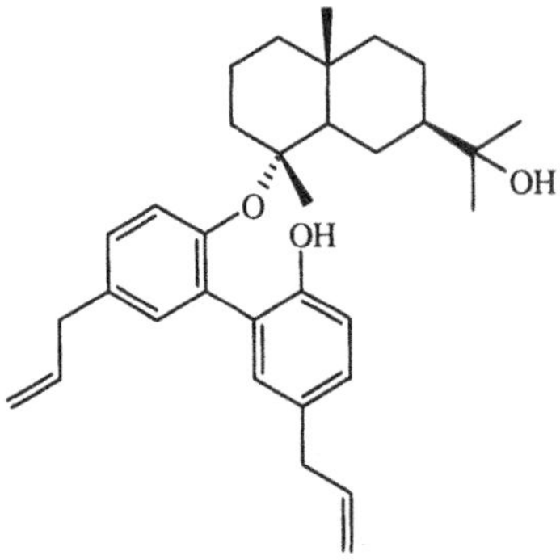   Eudesmagnolol (**133**)

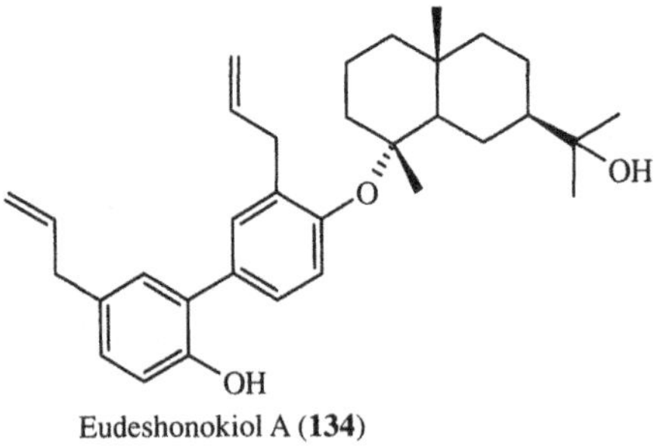

Eudeshonokiol A (**134**) 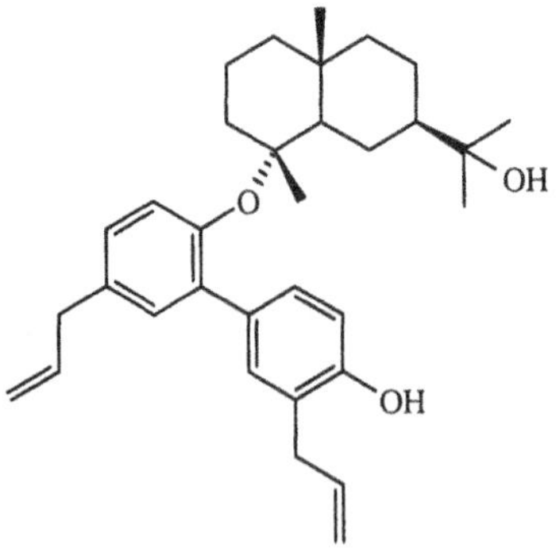   Eudeshonokiol B (**135**)

Figure 3.12 Terpenyl neolignans from *Magnolia*.

produces flavonoids, lignans and neolignans, it is not surprising that various phenylpropanoids (Figures 3.14 and 3.15) have also been isolated from at least nine different species of this genus. From *M. salicifolia* alone, seven such compounds —acetoside (**183**), *cis*- and *trans*-anethole (**167**, **168**), eugenol and eugenol methyl ether (**176**, **177**), isoeugenol and methylisoeugenol (**170**, **172**), 3,4-methylenedioxy-

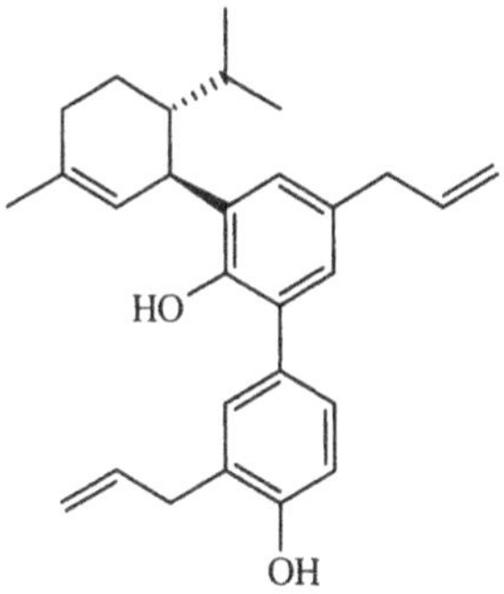

Piperitylhonokiol (**136**) 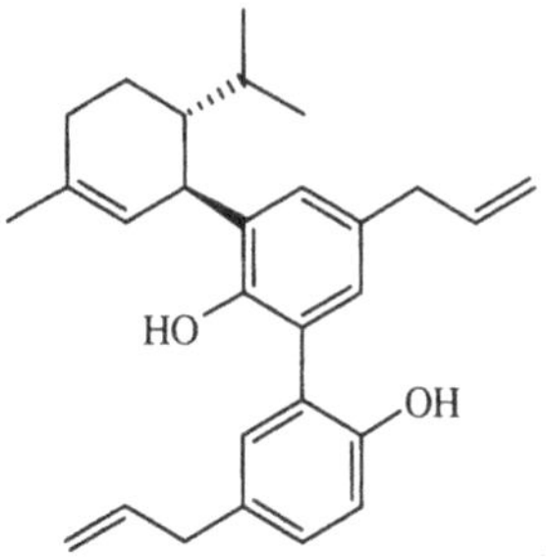   Piperitylmagnolol (**137**)

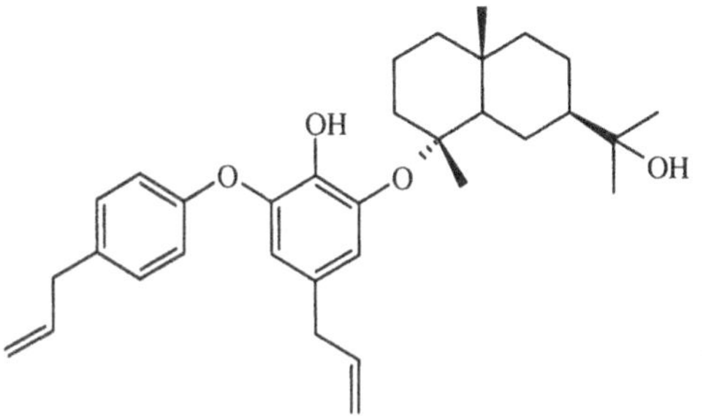

Eudesobovatol A (**138**) 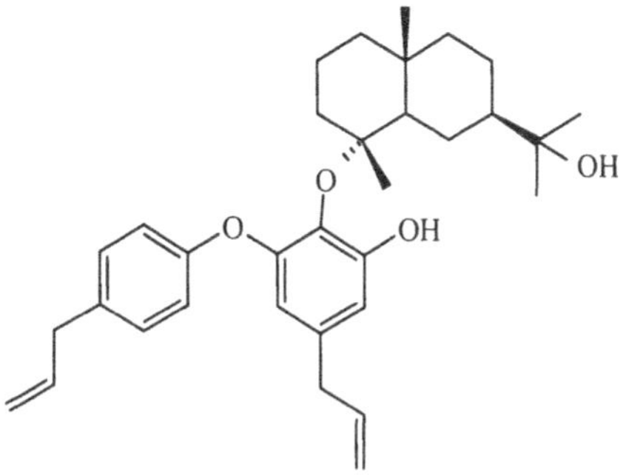   Eudesobovatol B (**139**)

Figure 3.12 (continued)

cinnamyl alcohol (173), 1-(2,4,5-trimethoxyphenyl)-1,2-propanedione (165) and saffrole (178), have been reported (Table 3.1). A total of 22 different phenylpropanoids, of which 14 are free (165–178) and eight are glycosylated (179–186), have been found in this genus. Most of them have more than one oxygenated substituent on the benzene ring. Glucose, rhamnose and galactose are the sugars normally found in phenylpropanoid glycosides produced by *Magnolia* and, except for coniferin (184) and syringin (185), more than one sugar is involved in glycoside formation. Syringin (185) appears to be the most common phenylpropanoid in *Magnolia* (Table 3.1).

3.2.7 Terpenoids

Terpenoids are naturally occurring compounds containing 10, 15, 20, 25, 30 or 40 carbon atoms and are formed from condensation and hydrogenation of two, three, four, five, six or eight isoprene units via the isopentyl diphosphate pathway. Like the acylpolymalonate pathway, the isopentyl diphosphate pathway starts with acetyl-CoA (Samuelsson, 1999). In *Magnolia*, 52 different terpenoids (187–239) have been isolated from various species (Figures 3.16 to 3.20). Except for four tetraterpenes (236–239), all these terpenoids are either monoterpenes or sequiterpenes. However, different classes of sequiterpene lactones are the major terpenoids in *Magnolia*. To our knowledge, no di- or triterpene has ever been reported from this genus.

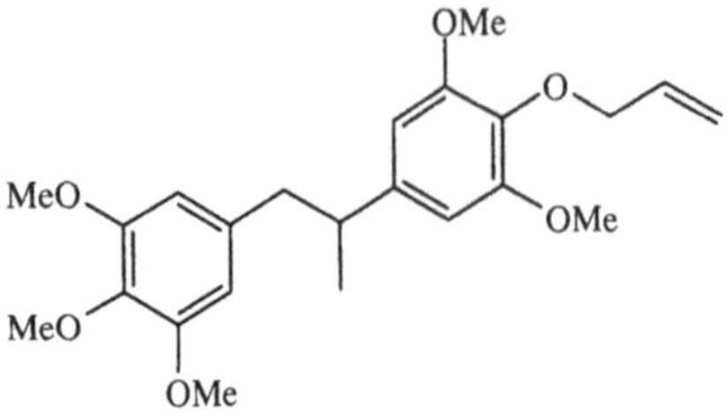

Aurein (**140**)

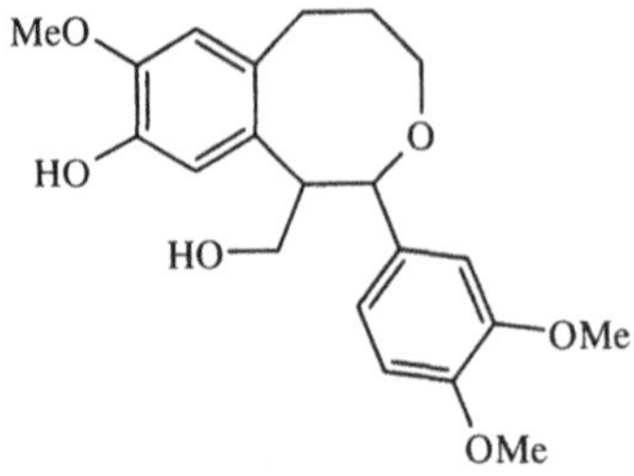

Biondinin A (**141**)

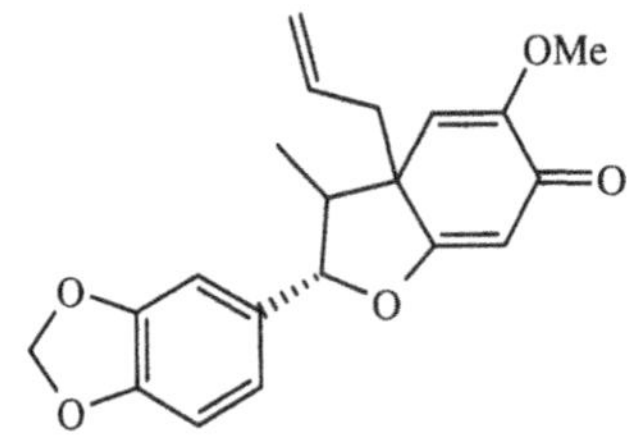

Burchellin (**142**)

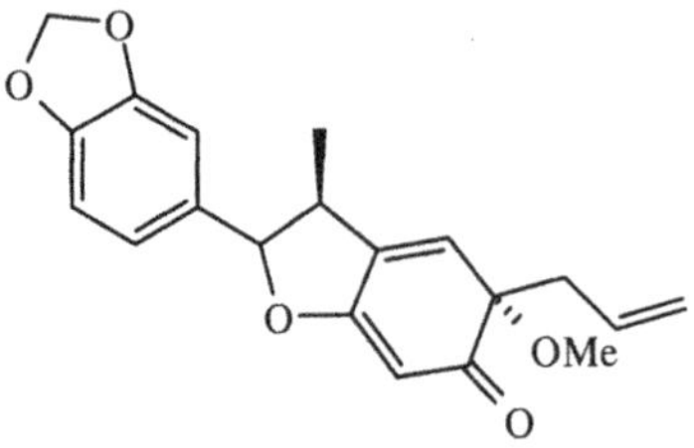

Cyclohexadienone (**143**)

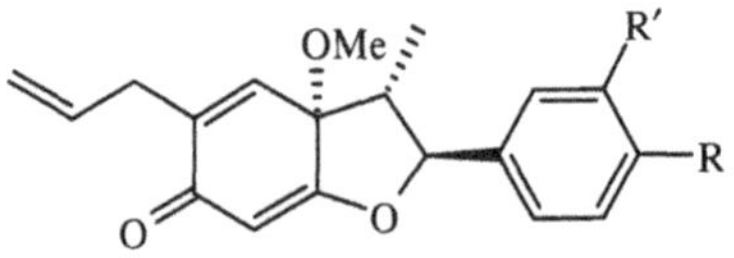

	R	R'
Denudatin A (**144**)	-O-CH$_2$-O-	
Denudatin B (**145**)	OMe	OMe

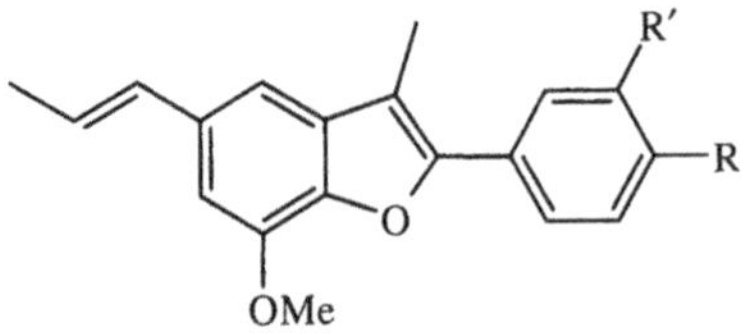

	R	R'
Eupomatenoid-1 (**146**)	-O-CH$_2$-O-	
Eupomatenoid-7 (**147**)	OH	OMe

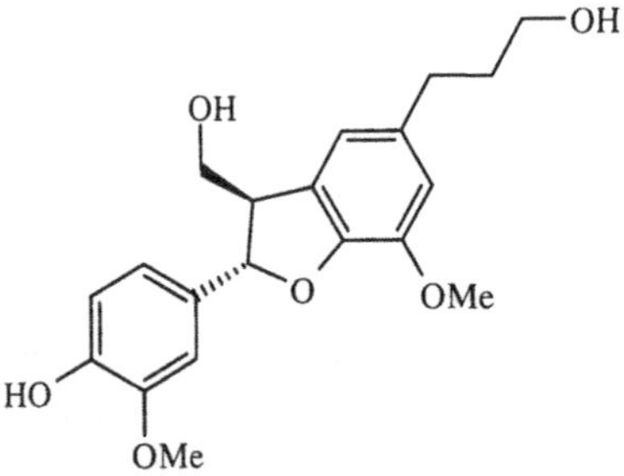

Dihydrodehydrodiconiferyl alcohol (**148**)

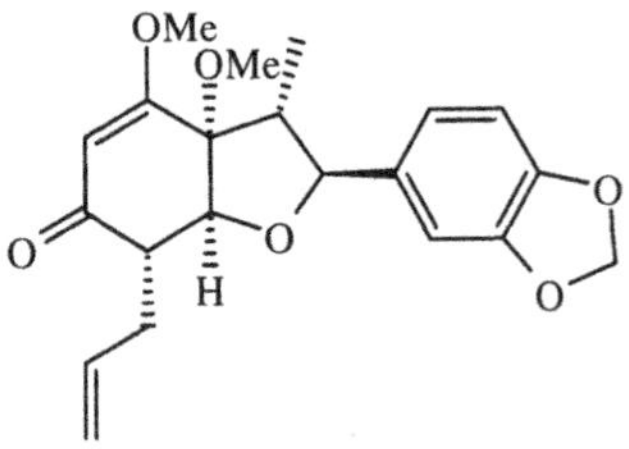

Fargesone A (**149**)

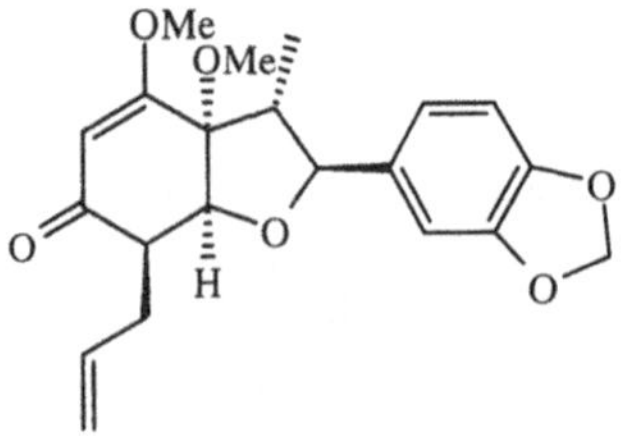

Fargesone B (**150**)

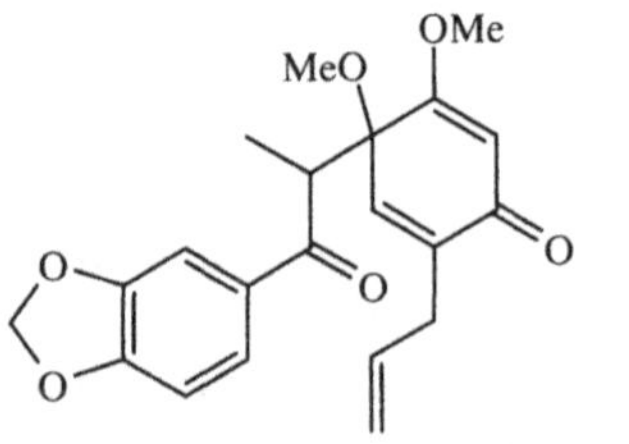

Fargesone C (**151**)

Figure 3.13 Other neolignans from *Magnolia*.

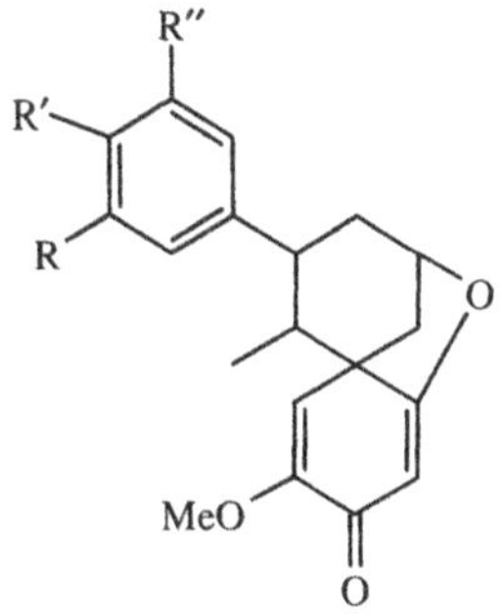

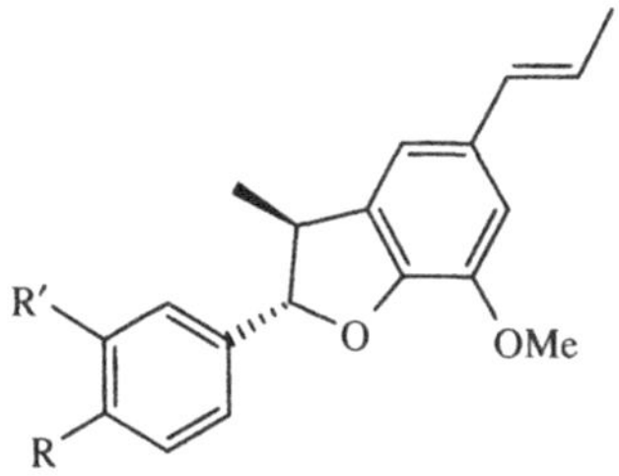

	R	R'	R''
Denudatone (**152**)	OMe	OMe	OMe
Futoenone (**153**)	-O-CH₂-O-		H

	R	R'
Licarin A (**154**)	OH	OMe
Licarin B (**155**)	-O-CH₂-O-	

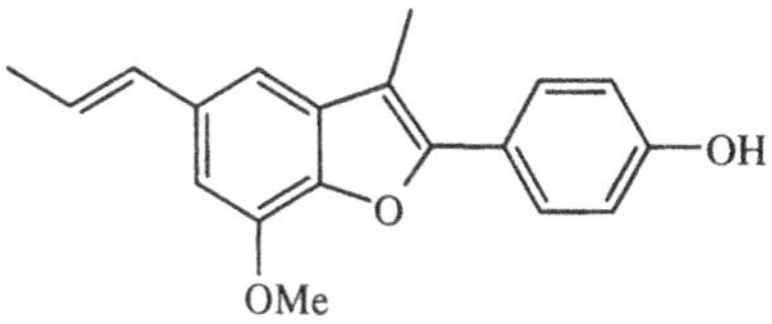

Kachirachirol A (**156**)

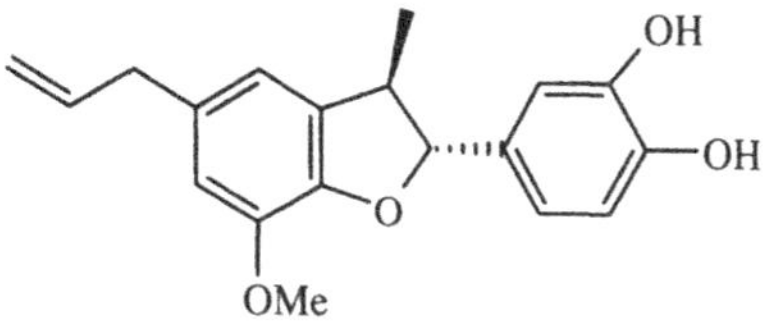

Kachirachirol B (**157**)

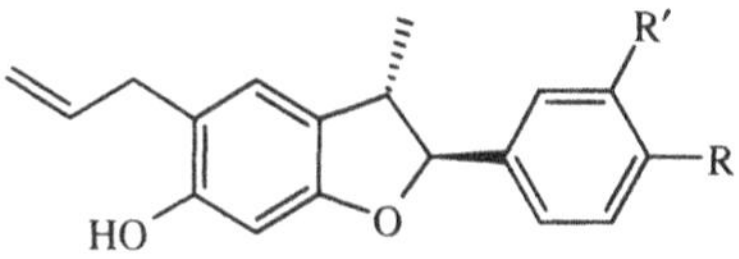

	R	R'
Liliflol A (**158**)	-O-CH₂-O-	
Liliflol B (**159**)	OMe	OMe

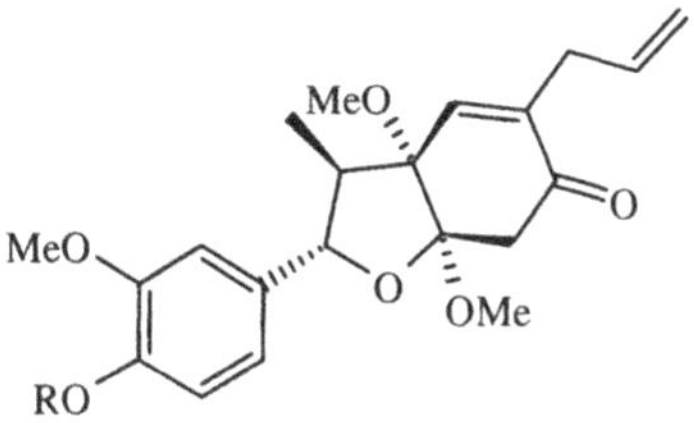

Liliflone (**160**)	R = H
Piperinone (**161**)	R = Me

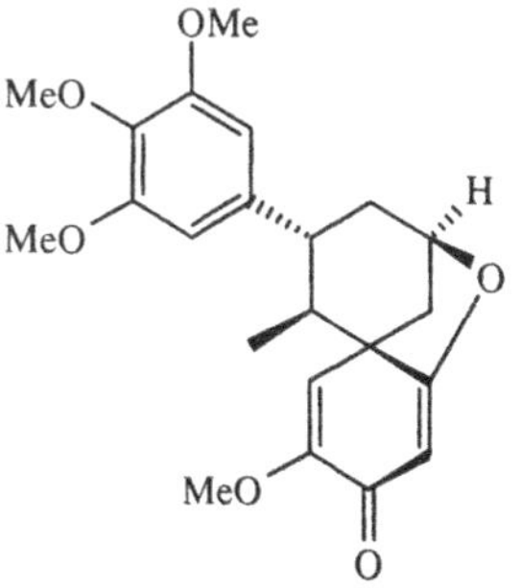

(−)-Maglifloenone (**162**)

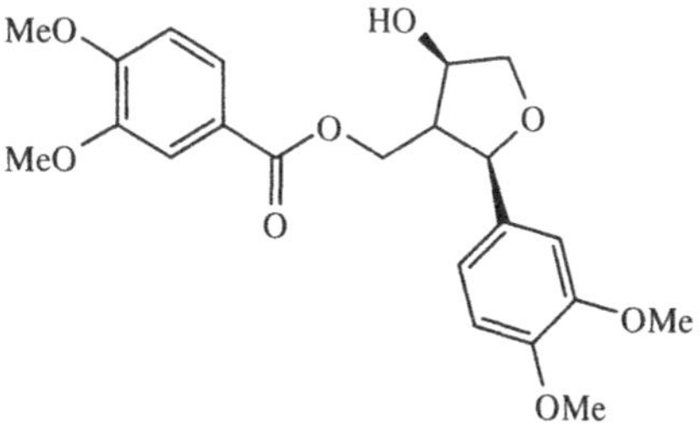

Magnostellin B (**163**)

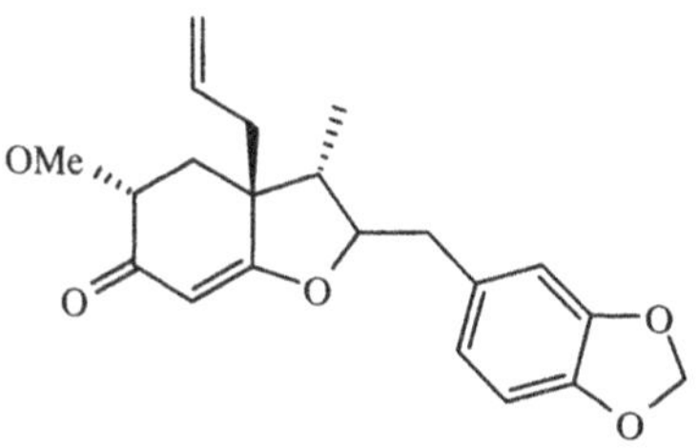

Soulangianin-I (**164**)

Figure 3.13 (continued)

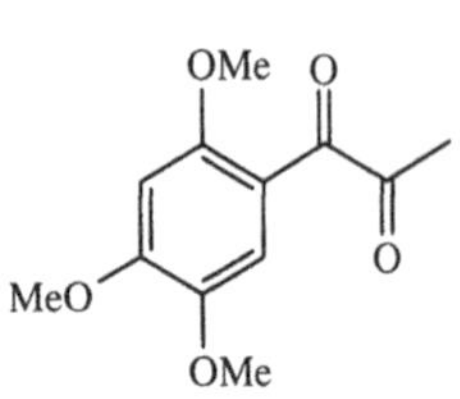

1-(2,4,5-Trimethoxyphenyl)-1,2-propanedione (**165**) 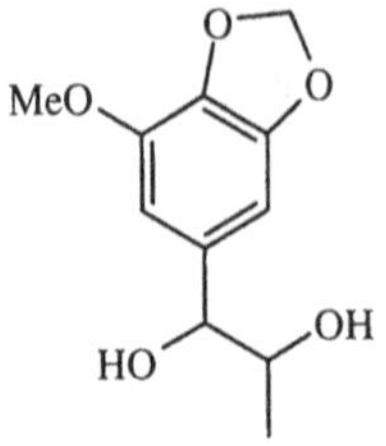   Diasyllaserine (**166**) 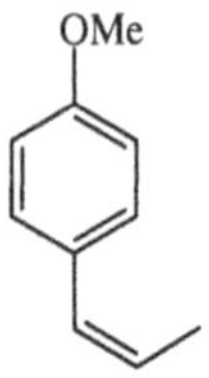   *cis*-Anethole (**167**)

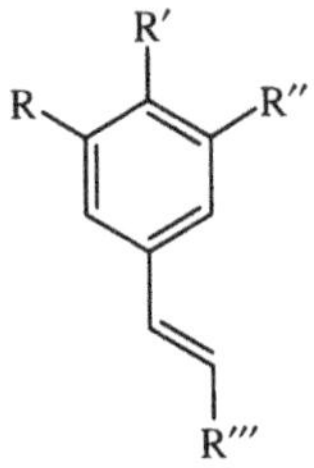

	R	R′	R″	R‴
trans-Anethole (**168**)	H	OMe	H	Me
4-Hydroxy-3-methoxycinnamic acid (**169**)	OMe	OH	H	COOH
Isoeugenol (**170**)	OMe	OH	H	Me
trans-Isomyristicine (**171**)	-O-CH₂-O-		OMe	Me
Methylisoeugenol (**172**)	OMe	OMe	H	Me
3,4-Methylenedioxycinnamyl alcohol (**173**)	-O-CH₂-O-		H	CH₂OH
Sinapyl alcohol (**174**)	OMe	OH	OMe	CH₂OH
Sinapic aldehyde (**175**)	OMe	OH	OMe	CHO

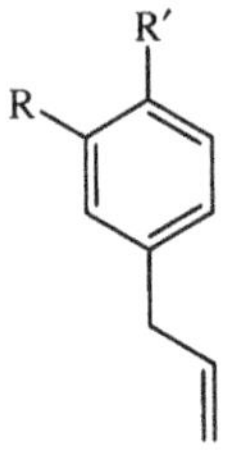

	R	R′
Eugenol (**176**)	OMe	OH
Eugenol methyl ether (**177**)	OMe	OMe
Saffrole (**178**)	-O-CH₂-O-	

Figure 3.14 Phenylpropanoids from *Magnolia*.

3.2.7.1   *Monoterpenes*

Monoterpenes are biosynthesised from two isoprene units, and thus contain a C_{10} skeleton, either as straight chain (190–193, 202) or as cyclic form (187–189, 194–201, 203) (Figure 3.16). Seventeen different monoterpenes have been reported from *M. biondii* (Han, 1993), *M. fargesii* (Kakisawa *et al.*, 1972), *M. obovata* (Fukuyama *et al.*, 1992) and *M. salicifolia* (Fujita and Fujita, 1974). *M. salicifolia* is the predominant source for this type of compound, and produces at least 15 monoterpenes (Table 3.1).

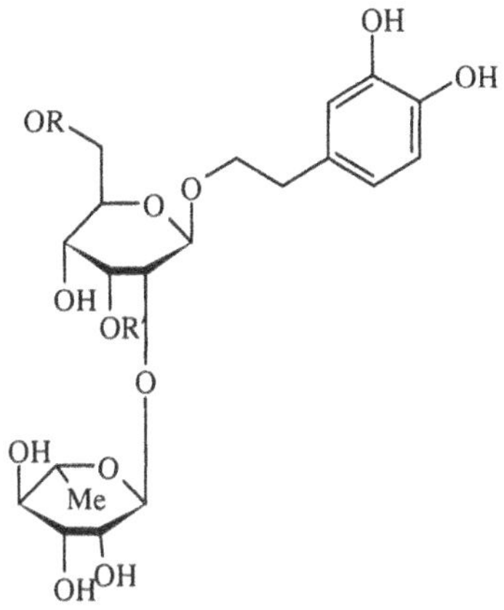

	R	R'		
Magnoloside A (**179**)	H	Caffeoyl	Magnoloside C (**181**)	R = Caffeoyl
Magnoloside B (**180**)	Glucosyl	Caffeoyl		

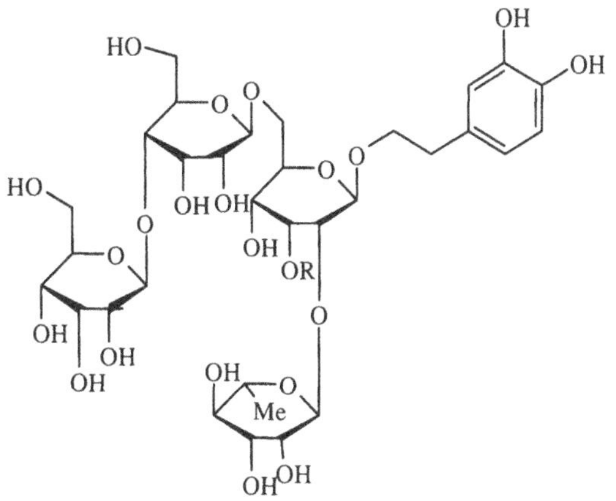

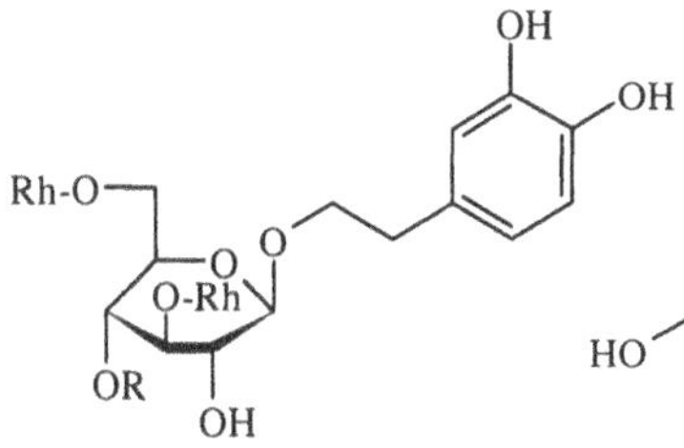

Magnolidin (**182**) R = Caffeoyl

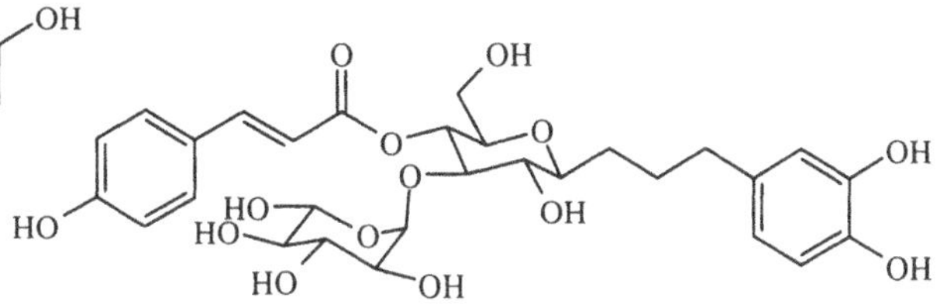

Acteoside (**183**)

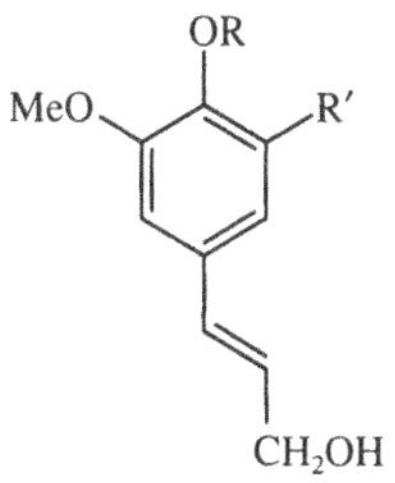

	R	R'
Coniferin (**184**)	Glu	H
Syringin (**185**)	Glu	OMe
Syringinoside (**186**)	Glu-Glu	OMe

Figure 3.15 Phenyl propanoid glycosides from *Magnolia*.

Citral A and B (**190**, **191**), cineol (**194**), camphor (**195**), α- and β-pinene (**196**, **197**), and camphene (**203**) are major monoterpenes found in *Magnolia*. Biondinin C and D (**187**, **188**), isolated from *M. biondii* (Han, 1993; DNP, 1999), possess a cinnamoyl moiety attached to the monoterpene unit (borneol).

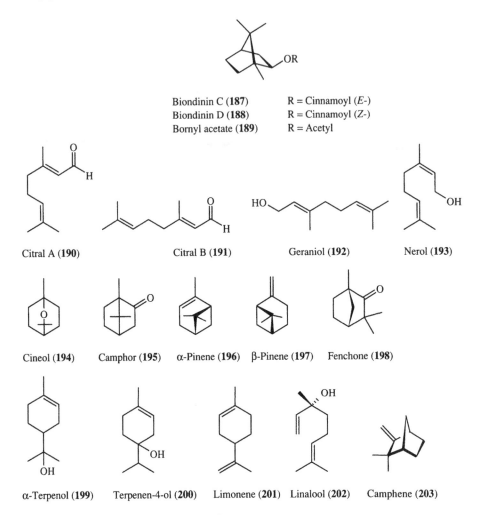

Biondinin C (**187**)   R = Cinnamoyl (*E-*)
Biondinin D (**188**)   R = Cinnamoyl (*Z-*)
Bornyl acetate (**189**)   R = Acetyl

Citral A (**190**)   Citral B (**191**)   Geraniol (**192**)   Nerol (**193**)

Cineol (**194**)   Camphor (**195**)   α-Pinene (**196**)   β-Pinene (**197**)   Fenchone (**198**)

α-Terpenol (**199**)   Terpenen-4-ol (**200**)   Limonene (**201**)   Linalool (**202**)   Camphene (**203**)

Figure 3.16 Monoterpenes from *Magnolia*.

3.2.7.2   *Sesquiterpenes*

Sesquiterpenes are C_{15} compounds formed by the assembly of three isoprene (C_5) units. In fact, geranyl diphosphate condenses further with one molecule of isopentyl diphosphate, leading to the formation of farnesyl diphosphate (FPP), which then undergoes cyclisation or skeletal rearrangements in several different ways, giving rise to a great number of sesquiterpenes (Samulelsson, 1999). A total of 32 secondary metabolites (**204–235**) belonging to this class (Figures 3.17 to 3.19), have been isolated from 14 different species of *Magnolia* (Table 3.1). Except for 9-oxonerolidol (**227**), 3-hydroxy-1,6,10-phytatrien-9-one (**228**), kobusimin A (**235**) and kobusimin B (**234**), all other sequiterpenes found in *Magnolia* are cyclic compounds. The majority of these sesquiterpenes can be described as costunolide/parthenolide type (**204–216**) and eudesmane-type (**217–223**). However, there are a few other types of sesquiterpenes also found in *Magnolia* (Figure 3.19).

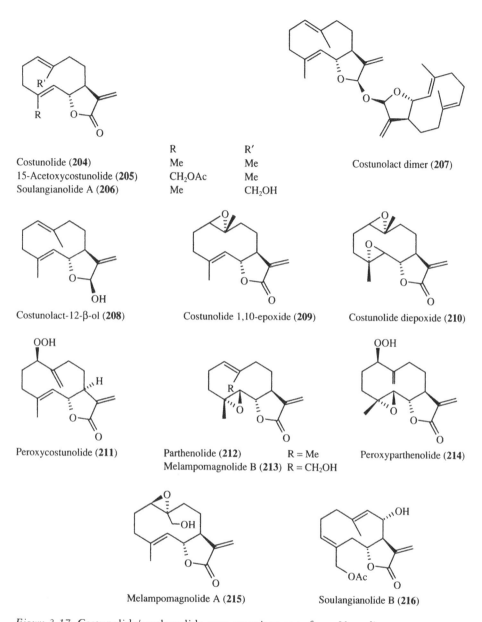

Costunolide (**204**)   R: Me   R′: Me
15-Acetoxycostunolide (**205**)   CH₂OAc   Me
Soulangianolide A (**206**)   Me   CH₂OH

	R	R′
Costunolide (**204**)	Me	Me
15-Acetoxycostunolide (**205**)	CH$_2$OAc	Me
Soulangianolide A (**206**)	Me	CH$_2$OH

Costunolact dimer (**207**)

Costunolact-12-β-ol (**208**)

Costunolide 1,10-epoxide (**209**)

Costunolide diepoxide (**210**)

Peroxycostunolide (**211**)

Parthenolide (**212**)   R = Me
Melampomagnolide B (**213**)   R = CH$_2$OH

Peroxyparthenolide (**214**)

Melampomagnolide A (**215**)

Soulangianolide B (**216**)

Figure 3.17 Costunolide/parthenolide-type sesquiterpenes from *Magnolia*.

3.2.7.2.1   COSTUNOLIDE/PARTHENOLIDE-TYPE SESQUITERPENES

This type is also known as 12,6-germacranolides. Costunolide (**204**) and parthenolide (**212**) are the two most commonly occurring sesquiterpenes in *Magnolia*, and have been isolated, respectively, from six and four different species (Table 3.1). Figure 3.17 shows all 13 sesquiterpenes of this type isolated from *Magnolia*. A costunolide dimer (**207**) was isolated from the leaves of *M. virginiana* (Song *et al.*, 1998). The presence of a hydroperoxy group on both costunolide and parthenolide skeletons has been

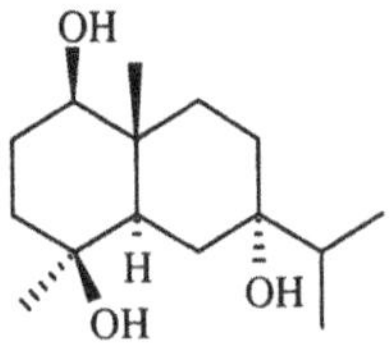

1β,4β,7α-Trihydroxyeudesmane (**217**) 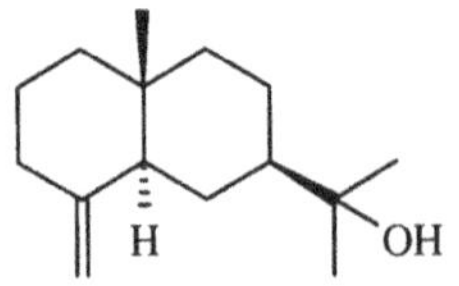   β-Eudesmol (**218**) 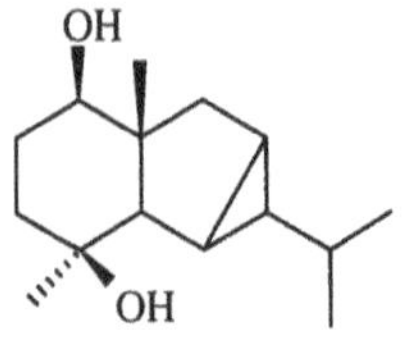   5α,7α(H)-6,8-cycloeudesma-
1β,4β-diol (**219**)

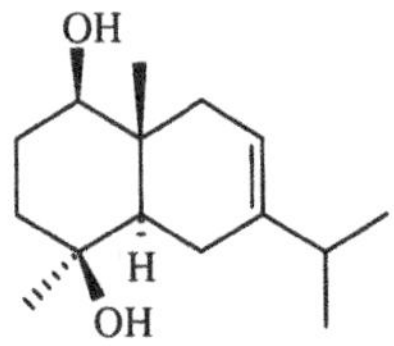

Oplodiol (**220**) 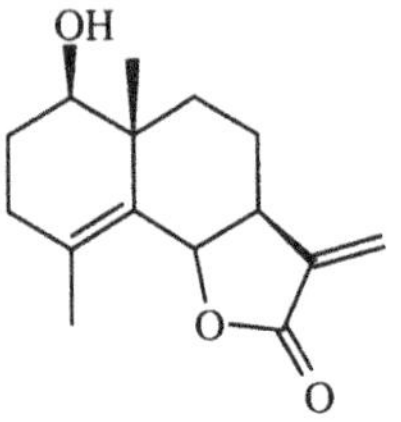   Magnolialide (**221**) 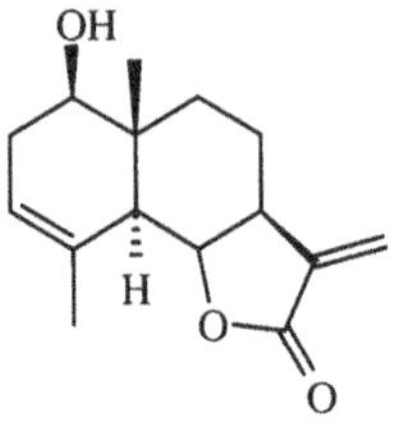   Santamarin (**222**) 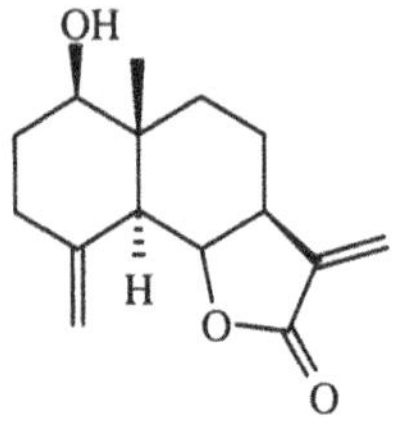   Reynosin (**223**)

Figure 3.18 Eudesmane-type sesquiterpenes from *Magnolia*.

observed in the compounds peroxycostunolide (**211**) and peroxyparthenolide (**214**), isolated from the leaves of *M. grandiflora* (El-Feraly *et al.*, 1977, 1979a).

3.2.7.2.2   EUDESMANE-TYPE SESQUITERPENES

Seven eudesmane-type sesquiterpenes (Figure 3.18) have been reported from *Magnolia*. β-Eudesmol has been found in three species, *M. obovata* (Mori *et al.*, 1997), *M. officinalis* (Konoshima *et al.*, 1991) and *M. rostrata* (Yan, 1979). Presence of a lactone ring is observed in manolialide (**221**), santamarin (**222**) and reynosin (**223**), isolated from root-barks of *M. grandiflora* (El-Feraly *et al.*, 1979b).

3.2.7.2.3   OTHER SESQUITERPENES

In addition to costunolide/parthenolide-type and eudesmane-type sesquiterpenes, *Magnolia* also produces 12 other sesquiterpenes (Figure 3.19). Four farnesane-type sesquiterpenes, 9-oxonerolidol (**227**) 3-hydroxy-1,6,10-phytatrien-9-one (**228**) kobusimin B (**234**) and kobusimin A (**235**), have been isolated from the leaves of *M. kobus* (Iida *et al.*, 1982a). Fukuyama *et al.* (1992) reported one caryophyllane-type compound, caryophyllene (**226**) from the leaves of *M. obovata*. An oppositane-type sesquiterpene, homalomenol (**225**), and an oplopane-type, oplopanone (**224**), were found in the flower buds of *M. fargesii* (Jung *et al.*, 1997). *M. watsonii* produces two guaiane-type sesquiterpenes, watsonol A (**229**) and watsonol B (**230**) (Ito *et al.*, 1984a,b). One aromadendrane-type, cycloclorenone (**233**) and a 12,6-guainolide-type sesquiterpene, magnograndiolide (**231**) were isolated, respectively, from the barks and leaves of *M. grandiflora* (Jacyno *et al.*, 1991; Halim *et al.*, 1984), and a guainolide-glucoside, trifloculoside (**232**) was found in the leaves of *M. virginiana* (Song *et al.*, 1998).

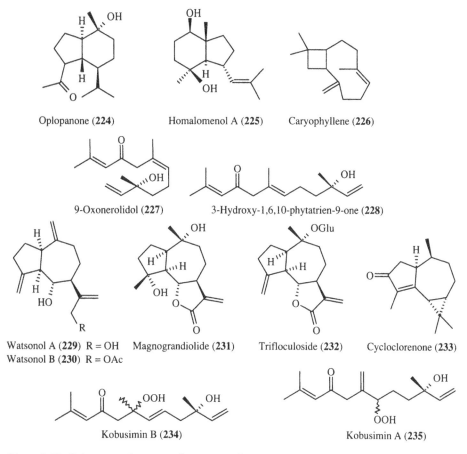

Oplopanone (**224**) Homalomenol A (**225**) Caryophyllene (**226**)

9-Oxonerolidol (**227**) 3-Hydroxy-1,6,10-phytatrien-9-one (**228**)

Watsonol A (**229**) R = OH Magnograndiolide (**231**) Trifloculoside (**232**) Cyclochlorenone (**233**)
Watsonol B (**230**) R = OAc

Kobusimin B (**234**) Kobusimin A (**235**)

Figure 3.19 Other sesquiterpenes from *Magnolia*.

3.2.7.3 Tetraterpenes

Only four tetraterpenes (C_{40} compounds), α-carotene 5,6-epoxide (**236**), β-carotene (**237**), lutein (**238**) and lutein 5,6-epoxide (**239**) have been found from *Magnolia* species (Figure 3.20).

3.2.8 Other secondary metabolites

In addition to the major classes of secondary metabolites (Figures 3.1 to 3.20) described above, 16 other compounds, most of which are simple phenol derivatives or benzenoids (240–246) have been reported from *Magnolia* (Figure 3.21). Syringaldehyde (**244**) and vanillin (**245**) are the two most common phenol derivatives found in this genus. Liliflodione (**255**), isolated from the leaves of *M. liliflora* (Iida and Ito, 1983) is an unusual type of compound. Chromanols (**250**) and α-tocopherol (**249**) have been found in *M. soulangeana* (Lichtenthaler, 1965) and vomifoliol (**253**) in *M. stellata* (Brieskorn and Huber, 1976).

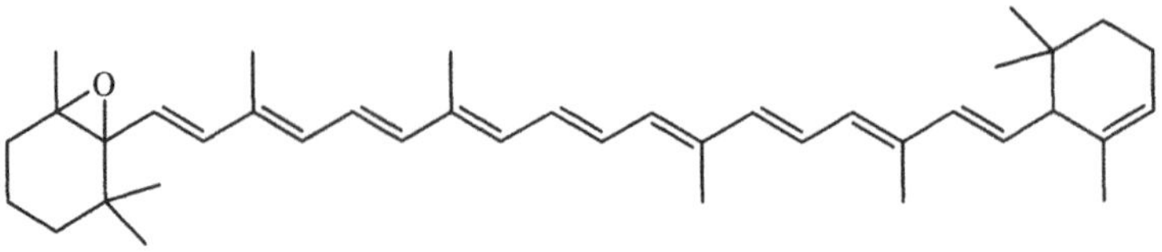

α-Carotene 5,6,-epoxide (**236**)

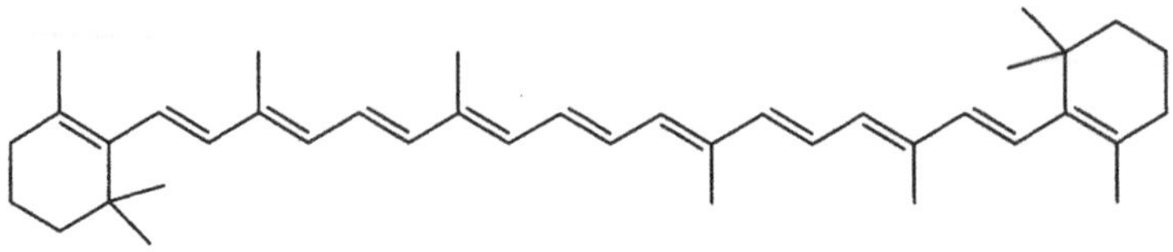

β-Carotene (**237**)

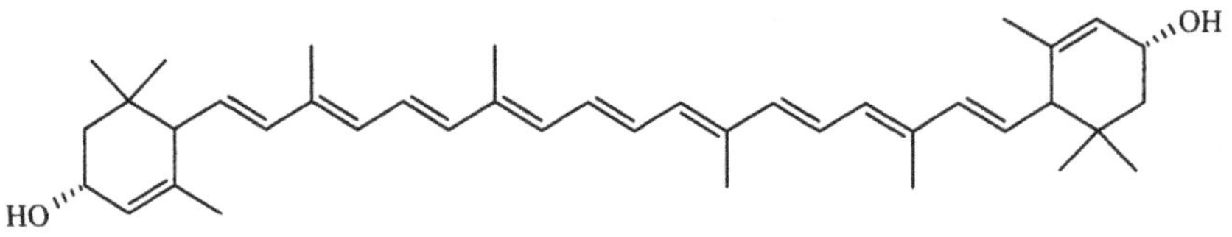

Lutein (**238**)

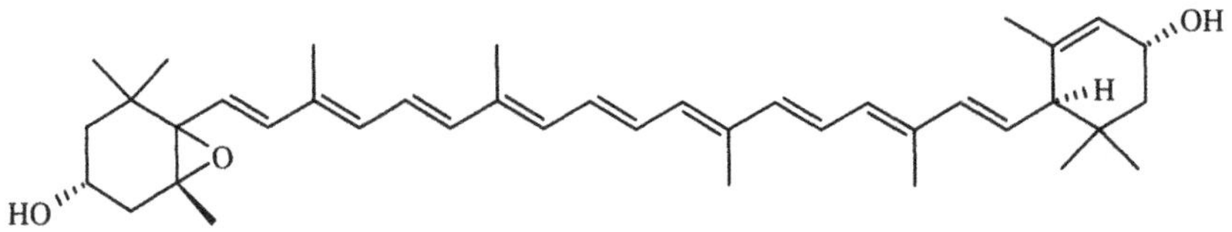

Lutein 5,6-epoxide (**239**)

Figure 3.20 Tetraterpenes from *Magnolia*.

3.3   Chemotaxonomic significance

Reports of phytochemical studies on less than 50% of the total number of *Magnolia* species are available to date. The amount and composition of classes of compounds such as alkaloids, flavonoids, essential oils and many others are governed by the age of the plant or its parts, the geographical source of the plants investigated, and their general habitat (Hegnauer, 1986). Therefore, detailed information on these factors and comprehensive data on chemical variations among the plants are essential for chemotaxonomic evaluation. On the basis of the data available from published results, it is somewhat difficult to draw any conclusion on the chemotaxonomy of the genus *Magnolia*. However, it is fair to say that the phytochemistry of *Magnolia* looks similar to that of other allied genera, especially *Liriodendron* and *Michelia*, within the family Magnoliaceae. The presence or absence of different classes of plant secondary metabolites in different species of this genus is summarised in Table 3.2. It can be noted that some species of *Magnolia* predominantly produce alkaloids, whereas some others produce lignans/neolignans as the major class of secondary metabolite (Table 3.1). Only *M. grandiflora* has been found to produce all classes of compounds described in this chapter. All other species are lacking in one or more classes.

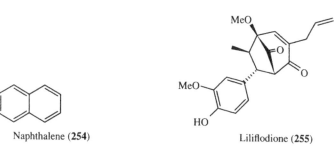

	R	R'	R''	R'''	R''''
Anisaldehyde (**240**)	CHO	H	H	OMe	H
Asaraldehyde (**241**)	CHO	OMe	H	OMe	OMe
p-Cymene (**242**)	Me	H	H	CH(CH$_3$)$_2$	H
Myristicine aldehyde (**243**)	CHO	H	OMe	-O-CH$_2$-O-	
Syringaldehyde (**244**)	CHO	H	OMe	OH	OMe
Vanillin (**245**)	CHO	H	OMe	OH	H
Vetratric acid (**246**)	COOH	H	OMe	OH	H

(*E*)-1,2,3,15-Tetranor-4,6,10-farnesatriene (**247**)

4,8,12-Trimethyl-1,3,7,11-tridecatetraene (**248**)

α-Tocopherol (**249**)

2-Chromanol (**250**)

Quinic acid (**251**)

Shikimic acid (**252**)

Vomifoliol (**253**)

Naphthalene (**254**)

Liliflodione (**255**)

Figure 3.21 Various other compounds from *Magnolia*.

Table 3.2 Distribution[a] of different classes of secondary metabolites in *Magnolia* species

Species	Alkaloids	Coumarins	Flavonoids	Lignans	Neolignans	Phenylpropanes	Terpenoids	Others
M. acuminata	+	−	−	+	+	−	−	−
M. asbei	−	−	+	+	−	−	+	−
M. biondii	−	−	+	+	+	−	−	−
M. campbellii	+	+	+	+	−	−	−	−
M. coco	+	−	−	−	−	−	−	−
M. compressa	+	−	+	+	+	−	+	−
M. cordata	−	−	+	−	−	+	+	+
M. denudata	+	−	+	−	+	+	−	−
M. fargesii	−	−	+	+	+	−	+	−
M. fraseri	−	−	−	−	−	−	−	+
M. fuscata	+	−	+	−	+	−	+	−
M. grandiflora	+	+	+	+	+	+	+	+
M. henryi	−	−	−	−	+	−	−	−
M. kachirachirai	+	−	+	+	+	−	+	−
M. kobus	+	−	+	−	−	−	−	+
M. lennei	−	−	−	+	−	−	+	−
M. liliifolia	+	−	+	−	+	−	−	−
M. liliiflora	+	+	+	+	−	−	−	+
M. macrophylla	−	−	+	−	−	−	+	−
M. mutabilis	+	−	−	+	+	+	+	−
M. obovata	+	−	−	+	+	+	−	+
M. officinalis	−	−	−	−	+	+	+	−
M. parviflora	−	−	−	+	−	−	+	−
M. pterocarpa	−	+	−	−	+	−	+	−
M. rostrata	−	−	+	+	−	−	−	+
M. salicifolia	+	−	+	−	+	+	+	−
M. sieboldii	+	−	+	+	−	+	−	+
M. sinensis	+	−	+	−	−	−	−	−
M. soulangeana	+	−	+	+	+	−	+	+
M. sprengeri	+	−	+	−	−	−	−	−
M. stellata	+	−	+	+	+	−	−	+
M. thompsoniana	−	−	−	−	−	−	−	−
M. tomentosa	+	−	+	−	+	−	−	+
M. tripetala	+	−	+	−	+	−	+	−
M. virginiana	−	−	+	−	+	−	+	+
M. watsonii	+	−	−	+	+	−	+	−
M. yulan	−	−	+	−	−	−	−	−

[a] + = present, − = absent

Alkaloids have been isolated from 23 species (Table 3.2), and of these *M. fuscata*, *M. parviflora* and *M. sprengeri* produce nothing but alkaloids. Aporphine/noraporphine-type and other isoquinoline alkaloids that are isolated from *Magnolia* have previously been used as chemotaxonomic markers for other plant families (Hegnauer, 1986) and are well-distributed among the plant families of the superorder Magnoliiflorae. Furmanowa and Jozefowicz (1980) used alkaloids as chemotaxonomic markers in some species of *Magnolia* and *Liriodendron*, and on the basis of their findings it was indicated that alkaloids could be useful as diagnostic traits in the comparative phytochemistry and chemotaxonomy of *Magnolia* species.

Coumarins have been reported from only four species, *M. compressa*, *M. grandiflora*, *M. macrophylla* and *M. pterocarpa*. All these coumarins are rather simple, widely distributed in the plant kingdom, and apparently do not have any chemotaxonomic significance, at least in the genus *Magnolia*.

Flavonoids have been found in 21 species, of which, *M. cordata*, *M. fraseri*, *M. lennei*, *M. sinensis*, *M. thompsonia*, and *M. yulan* seem to produce no other secondary metabolites apart from flavonoids. From the taxonomic viewpoint, anthocyanidins and other flavonoids found in *Magnolia* might be of some interest because of the variety of glycosidic combinations within the structures. The relatively complex structures of the anthocyanidins found in *Magnolia* suggest that the Magnoliaceae is not a primitive family but is relatively advanced in evolutionary terms (Francis and Harborne, 1966). However, it should be noted that *Magnolia* does not produce flavonoids with more advanced evolutionary features such as extensive methoxyl groups, prenyl substituents and further ring formation leading to pyranoflavanoids.

While lignans are found in 16 *Magnolia* species, neolignans have been isolated from 17 species. Lignans and neolignans are widely distributed in several plant families and have little chemotaxonomic value. However, magnolol/honokiol-type neolignans seem to be very much characteristic of *Magnolia* species.

Nine species of *Magnolia* produce phenylpropanoids but, owing to their widespread occurrence in nature, these do not seem to carry any chemotaxonomic value. Among the terpenoids found in 16 species of this genus, sesquiterpene lactones of different types might have some chemotaxonomic implications, as this class of compounds has previously been used successfully in chemosystematic analysis of other plant families (Emerenciano *et al.*, 1987). On the basis of the occurrence of sesquiterpenes, Song *et al.* (1998) suggested that the southern variety of *M. virginiana* is closely aligned with *M. grandiflora*, which typically produces, besides other sesquiterpene lactones, costunolide (204) and parthenolide (212) as common constituents.

Azuma *et al.* (1996) studied nine taxa of *Magnolia* and noticed that occurrence of naphthalene was very much restricted to the species *M. denudata*, *M. kobus*, *M. liliiflora* and *M. tomentosa* of the subgenus *Yulania*, and no naphthalene was found in *M. grandiflora*, *M. obovata* or *M. sieboldii*, all of which belong to the subgenus *Magnolia*. On the basis of this observation, it was suggested that the formation of naphthalene might be one of the physiological characters that distinguish the subgenus *Yulania*. More systematic and comprehensive phytochemical studies are necessary to understand the chemotaxonomic significance of the secondary metabolites in *Magnolia*.

References

Abdallah, O.M. (1993) Lignans in flower buds of *Magnolia saulangiana*. *Phytochemistry*, 34, 1185–1187.

Azuma, H., Toyota, M., Asakawa, Y. and Kawano, S. (1996) Napthalene—a constituent of *Magnolia* flowers. *Phytochemistry*, 42, 999–1004.

Baek, N.I., Jun, H.K., Lee, Y.H., Park, J.D., Kang, K.S. and Kim, S.I. (1992) A new dehydrodieugenol from *Magnolia officinalis*. *Planta Med.*, 58, 566–567.

Baures, P.W., Miski, M. and Eggleston, D.S. (1992) Structure of sesamin. *Acta Crystallogr. C, Cryst. Struct. Commun.*, 48, 574–576.

Boudet, A. (1973) Quinic and shikimic acids in woody angiosperms. *Phytochemistry*, 12, 363–370.

Brieskorn, C.H. and Huber, H. (1976) Vier Neue Lignane Aus *Aptosimum spinescens* Thunbg. *Tetrahedron Lett.*, 26, 2221–2224.

Chae, S.H., Kim, P.S., Cho, J.Y., Park, J.S., Lee, J.H., Yoo, E.S., Baik, K.U., Lee, J.S. and Park, M.H. (1998) Isolation and identification of inhibitory compounds on TNF-alpha production from *Magnolia fargesii*. *Arch. Pharmacol. Res.*, 21, 67–69.

Chandra, A. and Nair, M.G. (1995) Supercritical carbon dioxide extraction and quantification of bioactive neolignans from *Magnolia virginiana* flowers. *Planta Med.*, 61, 192–195.

Chen, C.C., Huang, Y.L., Chen, Y.P., Hsu, H.Y. and Kuo, Y.H. (1988) Three new neolignans, fragesone A, fragesone B and fragesone C from the flower buds of *Magnolia fargesii*. *Chem. Pharm. Bull.*, 36, 1791–1795.

Chou, C.Y.C., Tsai, T.H., Lin, M.F. and Chen, C.F. (1996) Simultaneous determination of honokiol and magnolol in *Magnolia officinalis* by capillary zone electrophoresis. *J. Liq. Chromatogr. Relat. Technol.*, 19, 1909–1915.

Clark, A.M., El-Feraly, F.S. and Li, W.-S. (1981) Antimicrobial activity of phenolic constituents of *Magnolia grandiflora* L. *J. Pharm. Sci.*, 70, 951–952.

Demuth, P. and Santamour, F.S. (1978) Carotenoid flower pigments in *Liriodendron* and *Magnolia*. *Bull. Torrey Bot. Club*, 105, 65–66.

DNP (1999) *Dictionary of Natural Products* on CD-ROM release 8:1. Boca Raton, FL: Chapman and Hall.

Doskotch, R.W. and Flom, M.S. (1972) Acuminatin, a new bis-phenylpropide from *Magnolia acuminata*. *Tetrahedron*, 28, 4711–4717.

El-Feraly, F.S. (1983) Novel melampolides from *Magnolia soulangiana* Lennei. *Phytochemistry*, 22, 2239–2241.

El-Feraly, F.S. (1984) Melampolides from *Magnolia grandiflora*. *Phytochemistry*, 23, 2372–2374.

El-Feraly, F.S. and Li, W.-S. (1978) Phenolic constituents of *Magnolia grandiflora* L. seeds. *Lloydia*, 41, 442–449.

El-Feraly, F.S., Chan, Y.-M., Fairchild, E.H. and Doskotch, R.W. (1977) Peroxycostunolide and peroxyparthenolide: two cytotoxic germacranolide hydroperoxides from *Magnolia grandiflora*. Structural revision of verlotorin and artemonin. *Tetrahedron Lett.*, 23, 1973–1976.

El-Feraly, F.S., Chan, Y.M., Capiton, G.A., Doskotch, R.W. and Fairchild, E.H. (1979a) Isolation and characterisation of peroxycostunolide (verlotorin) and peroxyparthenolide from *Magnolia grandiflora*. Carbon-13 nuclear magnetic spectroscopy of costunolide and related compounds. *J. Org. Chem.*, 44, 3952–3955.

El-Feraly, F.S., Chan, Y.M. and Benigni, D.A. (1979b) Magnolialide: a novel eudesmanolide from the root bark of *Magnolia grandiflora*. *Phytochemistry*, 18, 881–882.

Emerenciano, V.P., Ferreira, Z.S., Kaplan, M.A.C. and Gottileb, O.R. (1987) A chemotaxonomic analysis of tribes of *Asteraceae* involving sesquiterpene lactones and flavonoids. *Phytochemistry*, 26, 3103–3115.

Francis, F.J. and Harborne, J.B. (1966) Anthocyanins and flavonol glycosides of *Magnolia* flowers. *Proc. Am. Soc. Hort. Sci.*, 89, 657–665.

Fujita, M., Itokawa, H. and Sashida, Y. (1973) Compounds of *Magnolia obovata*. III. Occurrence of magnolol and honokiol in *M. obovata* and other allied plants. *Yakugaku Zasshi*, 93, 429–434.

Fujita, S. and Fujita, Y. (1974) Miscellaneous contributions to the essential oils of the plants from various territories. XXXIV. Comparative biochemical and chemotaxonomical studies of the essential oils of *Magnolia salicifolia*. II. *Chem. Pharm. Bull.*, 22, 707–709.

Fukushima, K., Taguchi, S., Matsui, N. and Yasuda, S. (1996) Heterogeneous distribution of magnolignol glucosides in the stems of *Magnolia kobus*. *Mokuzai Gakkaishi*, 42, 1029–1031.

Fukuyama, Y., Otoshi, Y., Kodama, M., Hasegawa, T., Okazaki, H. and Nagasawa, M. (1989) Novel neurotrophic sesquiterpene neolignans from *Magnolia obovata*. *Tetrahedron Lett.*, 30, 5907–5910.

Fukuyama, Y., Otoshi, Y., Kodama, M., Hasegawa, T. and Okazaki, H. (1990b) Structure of clovanemagnolol, a novel neurotrophic sesquiterpene neolignan from *Magnolia obovata*. *Tetrahedron Lett.*, 31, 4477–4480.

Fukuyama, Y., Otoshi, Y., Nakamura, K., Kodama, M., Sugawara, M. and Nagasawa, M. (1990a) Structures of eudesmagnolol and eudeshonokiol, novel sesquiterpene neolignans isolated from *Magnolia obovata*. *Chem. Lett.*, 2, 295–296.

Fukuyama, Y., Otoshi, Y., Miyoshi, K., Nakamura, K., Kodama, M., Nagasawa, M., Hasegawa, T., Okazaki, H. and Sugawara, M. (1992) Neurotrophic sesquiterpene-neolignans from *Magnolia obovata*—structure and neurotrophic activity. *Tetrahedron*, 48, 377–392.

Fukuyama, Y., Otoshi, Y., Miyoshi, K., Hasegawa, N., Kan, Y. and Kodama, M. (1993) Structure of magnolianin, a novel trilignan possessing potent 5-lipoxygenase inhibitory activity. *Tetrahedron Lett.*, 34, 1051–1054.

Funayama, S., Adachi, M., Aoyagi, T. and Nozoe, S. (1995) Cytocidal principles of *Magnolia denudata*. *Int. J. Pharmacognosy*, 33, 21–24.

Furmanowa, M. and Jozefowicz, J. (1980) Alkaloids as taxonomic markers in some species of *Magnolia* L. and *Liriodendron* L. *Acta Soc. Bot. Pol.*, 49, 527–535.

Guinaudeau, H., Leboeuf, M. and Cave, A. (1975) Aporphine alkaloids. *Lloydia*, 38, 275–338.

Halim, A.F., Mansour, E.S., Badria, F.A., Ziesche, J. and Bohlmann, F. (1984) A guaianolide from *Magnolia grandiflora*. *Phytochemistry*, 23, 914–915.

Han, G. (1993) Constituents of the flowers of *Magnolia biondii*. *Cin. Chem. Lett.*, 4, 33–35.

Hasegawa, T., Fukuyama, Y., Yamada, T. and Nakagawa, K. (1988) Isolation and structure of magnoloside A, a new phenylpropanoid glycoside from *Magnolia obovata* Thunb. *Chem. Lett.*, 1, 163–166.

Hayashi, K. and Ouchi, K. (1948) Colouring matter contained in the white flower of *Magnolia kobus*. *Proc. Jpn. Acad.*, 24, 16–19.

Hayashi, K. and Ouchi, K. (1949a) Plant pigments. III. Rutin from the flowers of *Magnolia kobus*. *Misc. Repts. Res. Inst. Nat. Resources (Japan)*, 14, 1–4.

Hayashi, K. and Ouchi, K. (1949b) Colouring matter in the white flowers of *Magnolia kobus*. *Acta Phytochim. (Japan)*, 15, 49–52.

Hegnauer, R. (1986) Phytochemistry and plant taxonomy—an essay on the chemotaxonomy of higher plants. *Phytochemistry*, 25, 1519–1535.

Hirose, M., Satoh, Y. and Hagitani, A. (1968) Lignans from the *Magnolia kobus* seeds. *Nippon Kagaku Zasshi*, 89, 889–891.

Huang, Y.L., Chen, C.C., Chen, Y.P., Hsu, H.Y. and Kuo, Y.H. (1990) (–)-Fargesol, a new lignan from the flower buds of *Magnolia fargesii*. *Planta Med.*, 56, 237–238.

Hufford, C.D. (1976) Four new *N*-acetylnoraporphine alkaloids from *Liriodendron tulipifera*. *Phytochemistry*, 15, 1169–1171.

Iida, T. and Ito, K. (1983) Four phenolic neolignans from *Magnolia liliflora*. *Phytochemistry*, 22, 763–766.

Iida, T., Nakano, M. and Ito, K. (1982a) Hydroperoxysesquiterpene and lignan constituents of *Magnolia kobus*. *Phytochemistry*, 21, 673–675.

Iida, T., Ichino, K. and Ito, K. (1982b) Neolignans from *Magnolia denudata*. *Phytochemistry*, 21, 2939–2941.

Iida, T., Noro, Y. and Ito, K. (1983) Magnostellin A and magnostellin B, novel lignans from *Magnolia stellata*. *Phytochemistry*, 22, 211–213.

Ito, K. and Asai, S. (1974) Alkaloids of magnoliaceous plants XL. Alkaloids of *Magnolia obovata*. 3. Bases of leaves and roots. *Yakugaku Zasshi*, 94, 729–734.

Ito, K., Iida, T., Ichino, K., Tsunezuka, M., Hattori, M. and Namba, T. (1982) Obovatol and obovatal, novel biphenyl ether lignans from the leaves of *Magnolia obovata* Thunb. *Chem. Pharm. Bull.*, 30, 3347–3353.

Ito, K., Iida, T. and Kobayashi, T. (1984a) Guaine sesquiterpenes from *Magnolia watsonii*. *Phytochemistry*, 23, 188–190.

Ito, K., Ichino, K., Iida, T. and Lai, J. (1984b) Neolignans from *Magnolia kachirachirai*. *Phytochemistry*, 23, 2643–2645.

Jacyno, J.M., Montemurro, N., Bates, A.D. and Cutler, H.G. (1991) Phytotoxic and antimicrobial properties of cyclocolorenone from *Magnolia grandiflora* L. *J. Agr. Food Chem.*, 39, 1166–1168.

Juneau, R.J. (1972) Mognolidine: new glycoside from the bark of *Magnolia grandiflora*. *Diss. Abstr. Int.*, B33, 3013.

Jung, K.Y., Kim, D.S., Oh, S.R., Lee, I.S., Lee, J.J., Lee H.K., Shin, D.H., Kim, E.H. and Cheong, C.J. (1997) Sesquiterpene components from the flower buds of *Magnolia fargesii*. *Arch. Pharmacol. Res.*, 20, 363–367.

Jung, K.Y., Oh, S.R., Park, S.H., Lee, I.S., Ahn, K.S., Lee, J.J. and Lee, H.K. (1998a) Anti-complement avtivity of tiliroside from the flower buds of *Magnolia fargesii*. *Biol. Pharm. Bull.*, 21, 1077–1078.

Jung, K.Y., Kim, D.S., Park, S.H., Lee, I.S., Oh, S.R., Lee, J.J., Kim, E.H., Cheong, C. and Lee, H.K. (1998b) 5α,7α(H)-6,8-Cycloeudesma-1β,4β-diol from the flower buds of *Magnolia fargesii*. *Phytochemistry*, 48, 1383–1386.

Jung, K.Y., Kim, D.S., Oh, S.R., Park, S.H., Lee, I.S., Lee, J.J., Shin, D.H. and Lee, H.K. (1998c) Magnone A and B, novel anti-PAF tetrahydrofuran lignans from the flower buds of *Magnolia fargesii*. *J. Nat. Prod.*, 61, 808–811.

Kakisawa, H., Chen, Y.P. and Hsu, H.Y. (1972) Lignans in flower buds of *Magnolia fargesii*. *Phytochemistry*, 11, 2289–2293.

Kamikado, T., Chang, C.F., Murakoshi, S., Sakurai, A. and Tamura, S. (1975) Isolation and structure elucidation of growth inhibitors on silkworm larvae from *Magnolia kobus* DC. *Agric. Biol. Chem.*, 39, 833–836.

Kapadia, G.J., Baldwing, H.H. and Shah, N.J. (1964a) Paper chromatography and identification of *Magnolia acuminata* alkaloids. *J. Pharm. Pharmacol.*, 16, 283–284.

Kapadia, G.J., Shah, N.J. and Highet, R.J. (1964b) Characterisation of a new Magnoliaceae alkaloid. *J. Pharm. Sci.*, 53, 1140–1141.

Kelm, M.A., Nair, M.G. and Schutzki, R.A. (1997) Mosquitocidal compounds from *Magnolia salicifolia*. *Int. J. Pharmacognosy*, 35, 84–90.

Kijjoa, A., Pinto, M.M.M., Tantisewie, B. and Herz, W. (1989) A biphenyl type neolignan and a biphenyl ether from *Magnolia henryi*. *Phytochemistry*, 28, 1284–1286.

Kikuchi, T., Kadota, S., Yanada, K., Tanaka, K. and Watanabe, K. (1983) Isolation and structure of magnosalin and magnoshinin, new neolignans from *Magnolia saliciforia* Maxim. *Chem. Pharm. Bull.*, 31, 1112–1114.

Komissarova, E.S. (1945) Action of *Magnolia fuscata* alkaloids on the cardiovascular system. *Farmacol. Toksikol.*, 8, 17–21.

Konoshima, T., Kozuka, M., Tokuda, H., Nishino, H., Iwashima, A., Haruna, M., Ito, K. and Tanabe, M. (1991) Studies on inhibitors of skin tunor promotion. 9. Neolignans from *Magnolia officinalis*. *J. Nat. Prod.*, 54, 816–822.

Kubo, I. and Yokokawa, Y. (1992) Two tyrosinase inhibiting flavonol glycosides from *Buddleia coriacea*. *Phytochemistry*, 31, 1075–1077.

Kwon, B.H., Kim, M.K., Lee, S.H., Kim, J.A., Lee, I.R., Kim, Y.K. and Bok, S.H. (1997) Acyl-CoA: cholesterol acyltransferase inhibitors from *Magnolia obovata*. *Planta Med.*, 63, 550–551.

Kwon, B.H., Jung, H.J., Lim, J.H., Kim, Y.S., Kim, M.K., Kim, Y.Mk., Bok, S.H., Bae, K.H. and Lee, I.R. (1999) Acyl-CoA: cholesterol acyltransferase inhibitory activity of lignans isolated from *Schizandra, Machilus* and *Magnolia* species. *Planta Med.*, 65, 74–76.

Li, W.-S. and El-Feraly, F.S. (1981) Studies on the Chemical components of the leaves of *Magnolia kachirachirai* Dandy. *Proc. Natl. Sci. Counc. Repub. China*, 5, 145–149.

Lichtenthaler, H.K. (1965) Isolation and characterisation of lipophilic phenol and chromanols from *Magnolia soulangeana*. *Z. Pflanzenphysiol.*, 53, 388–403.

Ma, Y.L., Huang, Q. and Han, G.Q. (1996) A neolignan and lignans from *Magnolia biondii*. *Phytochemistry*, 41, 287–288.

Massanet, G.M., Pando, E., Rodriguez-Luis, F. and Zubia, E. (1988) Lignans: a review. *Fitoterapia*, **LX**, 3–35.

Matsutani, H. and Shiba, T. (1975) Tyramine from *Magnolia* species. *Phytochemistry*, 14, 1132–1133.

Miyauchi, T. and Ozawa, S. (1998) Formation of (+)-eudesmin in *Magnolia kobus* DC. var. *Borealis* sarg. *Phytochemistry*, 47, 665–670.

Miyazawa, M., Kasahara, H. and Kameoka, H. (1992) Phenolic lignans from flower buds of *Magnolia fargesii*. *Phytochemistry*, 31, 3666–3668.

Miyazawa, M., Ishikawa, Y., Kasahara, H., Yamanaka, J. and Kameoka, H. (1994) An insect growth-inhibitory lignan from flower buds of *Magnolia fargesii*. *Phytochemistry*, 35, 611–613.

Miyazawa, M., Kasahara, H. and Kameoka, H. (1995) Absolute configuration of (−)-magnofargesin from *Magnolia fargesii*. *Nat. Prod. Lett.*, 7, 205–207.

Miyazawa, M., Kasahara, H. and Kameoka, H. (1996) (−)-Magnofargesin and (+)-magnoliadiol, two lignans from *Magnolia fargesii*. *Phytochemistry*, 42, 531–533.

Mori, M., Komatsu, M., Kido, M. and Nakagawa, K. (1987) Synthesis of lignans. 2. A simple biogenetic type synthesis of magnosalicin—a new neolignan with antiallergy activity isolated from *Magnolia salicifolia*. *Tetrahedron*, 42, 523–528.

Mori, M., Aoyama, M. and Doi, S. (1997) Antifungal constituents in the bark of *Magnolia obovata* Thunb. *Holz. Als. Roh-Und Werkstoff*, 55, 275–278.

Moriyasu, M. (1996) Alkaloid from *Magnolia officinalis*. *Nat. Med. (Tokyo)*, 50, 413–415.

Nagashima, S., Komiya, T., Murata, Y. and Masuoka, T. (1981) Antihistamine activity of Shin-i. *Takeda Kenkyushoho*, 40, 27–36.

Nakano, T. (1953) The alkaloids of magnoliaceous plants. IX. Alkaloids of *Magnolia lilifolia*. *Pharm. Bull. (Japan)*, 1, 29–32.

Nakano, T. (1954a) The alkaloids of magnoliaceous plants. XII. Alkaloids of *Magnolia grandiflora* (3). Structure of magnoflorine. *Pharm. Bull. (Japan)*, 2, 329–334.

Nakano, T. (1954b) Studies on the alkaloids of magnoliaceous plants. XIV. Alkaloids of *Magnolia grandiflora* L. *Pharm. Bull. (Japan)*, 2, 321–325.

Nakano, T. (1956) Alkaloids of magnoliaceous plants. XVI. Alkaloids of *Magnolia denudata* Desr. (2). *Pharm. Bull. (Japan)*, 4, 67–68.

Nakano, T. and Uchiyama, M. (1956a) Alkaloids of magnoliaceous plants. XVII. Alkaloids of *Magnolia parviflora* Sieb. et Zucc. *Pharm. Bull. (Tokyo)*, 4, 408–409.

Nakano, T. and Uchiyama, M. (1956b) Alkaloids of magnoliaceous plants. XVIII. Alkaloids of *Magnolia kobus* DC. var. *borealis* Koidz. *Pharm. Bull. (Tokyo)*, 409–410.

Namba, T. (1980) *Coloured Illustrations of Wakan-Yaku (The Crude Drugs in Japan, China and Neighbouring Countries)*, vol. II, pp. 127–129. Osaka: Hoikusha Publishing.

Nitao, J.K., Nair, M.G., Thorogood, D.L., Johnson, K.S. and Scriber, J.M. (1991) Bioactive neolignans from the leaves of *Magnolia virginiana*. *Phytochemistry*, 30, 2193–2195.

Ogiu, K. and Morita, M. (1953) Curane-like action of magnocurarine isolated from *Magnolia obovata*. *Jpn. J. Pharmacol.*, 2, 89–96.

Pan, J.X., Hensens, O.D., Zink, D.L., Chang, M.N. and Hwang, S.B. (1987) Lignans with platelet activating factor antagonist activity from *Magnolia biondii. Phytochemistry*, 26, 1377–1379.

Park, H.J., Jung, W.T., Basnet, P., Kadota, S. and Namba, T. (1996) Syringin 4-*O*-β-glucoside, a new phenylpropanoid glycoside, and costunolide, a nitric oxide synthase inhibitor, from the stem bark of *Magnolia sieboldii. J. Nat. Prod.*, 59, 1128–1130.

Park, J.B., Lee, C.K. and Park, H.J. (1997) Anti-*Helicobacter pylori* effect of costunolide isolated from the stem bark of *Magnolia sieboldii. Arch. Pharmacol Res.*, 20, 275–279.

Plouvier, V. (1943) The presence of rutoside in the flowers of certain *Magnolia. Compt. Rend.*, 216, 459–461.

Plouvier, V. (1962) The occurrence of syringoside in a number of botanical families. *Compt. Rend.*, 254, 4196–4198.

Plouvier, V. (1968) Coumarin heterosides: magnolioside from *Magnolia macrophylla* and calycanthoside and chicorioside from *Fraxinus. C. R. Acad. Aci. Paris. Ser. D.*, 266, 1526–1528.

Proskurnina, N.F. (1946) Alkaloids of magnolia fuscata III. Structure of magnolamine. *J. Gen. Chem. (USSR)*, 16, 129–138.

Proskurnina, N.F. and Orekhov, A.P. (1938) Alkaloids of *Magnolia fuscata. Bull. Soc. Chim.*, 5, 1357–1360.

Proskurnina, N.F. and Orekhov, A.P. (1940) Alkaloids of *Magnolia fuscata* II. Structure of magnoline. *J. Gen. Chem. (USSR)*, 10, 707–713.

Rao, K.V. (1975) Glycosides of *Magnolia grandiflora*. I. Isolation of three crystalline glycosides. *Planta Med.*, 27, 31–36.

Rao, K.V. and Davis, T.L. (1982a) Constituents of *Magnolia grandiflora*. Cyclocolorenone. *Planta Med.*, 44, 249–250.

Rao, K.V. and Davis, T.L. (1982b) Constituents of *Magnolia grandiflora*. 1. Mono-*O*-methylhonokiol. *Planta Med.*, 45, 57–59.

Rao, K.V. and Juneau, R.J. (1975) Glycosides of magnolia. II. Structure elucidation of magnolidin. *Lloydia*, 38, 339–342.

Rao, K.V. and Wu, W.-N. (1978) Glycosides of *Magnolia* III. Structural elucidation of magnolenin C. *Lloydia*, 41, 56–62.

Samuelsson, G. (1999) *Drugs of Natural Origin: A Textbook of Pharmacognosy*, 4th revised edn. Stockholm: Swedish Pharmaceutical Press.

Santamour, F.S., Jr (1965a) Biochemical studies in *Magnolia*. I. Floral anthocyanins. *Morris Arboretum Bull.*, 16, 43–48.

Santamour, F.S., Jr (1965b) Biochemical studies in *Magnolia*. II. Leucoanthocyanins in leaves. *Morris Arboretum Bull.*, 16, 63–64.

Santamour, F.S., Jr (1966a) Biochemical studies in *Magnolia*. III. Fruit anthocyanins. *Morris Arboretum Bull.*, 17, 13.

Santamour, F.S., Jr (1966b) Biochemical studies in *Magnolia*. IV. Flavonols and flavones. *Morris Arboretum Bull.*, 17, 65–68.

Sarker, S.D. (1997) Biological activity of magnolol: a review. *Fitoterapia*, LXVIII, 3–8.

Sashida, Y., Sugiyama, R., Iwasaki, S., Shimomura, H., Itokawa, H. and Fujita, M. (1976) Studies on the components of *Magnolia obovata* Thunb. V. Neutral and acidic components of the heartwood. *Yakugaku Zasshi*, 96, 659–662.

Sazaki, T. (1921) Alkaloids from *Magnolia obovata. Fukuoka Med. J.*, 14, 391.

Shinoda, Y., Nonomura, S. and Kawamura, I. (1976) Chemical composition of the family Magnoliaceae. *Gifu Daigaku Nogakubu Kenkyu Hokoku*, 39, 87–92.

Song, Q., GomezBarrios, M.L., Fronczek, F.R., Vargas, D., Thien, L.B. and Fischer, N.H. (1998) Sesquiterpenes from southern *Magnolia virginiana. Phytochemistry*, 47, 221–226.

Sugi, Y. (1930) Constituents of the bark of *Magnolia officinalis* Rhed. et Wils and *Magnolia obovata* Thumb. *J. Pharm. Sci. Jpn.*, 50, 183–217.

Tada, H., Fujioka, R. and Takayama, Y. (1982) 15-Acetoxycostunolide from *Magnolia sieboldii*. *Phytochemistry*, 21, 458–459.

Talpatra, B., Mukhopadhyay, P. and Datta, L.N. (1975) Alkaloids of *Magnolia campbellii* and *Magnolia mutabilis*. *Phytochemistry*, 14, 589–590.

Talpatra, B., Chaudhuri, P.K. and Talpatra, S.K. (1982) (–)-Maglifloenone, a novel spiro-cyclohexadienone neolignan and other constituents from *Magnolia liliflora*. *Phytochemistry*, 21, 747–750.

Talpatra, B., Ray, G. and Talpatra, S.K. (1983) Polyphenolic constituents of *Magnolia pterocarpa* Roxb. *J. Indian Chem. Soc.*, 60, 96–98.

Teng, C.M., Yu, S.M., Chen, C.C., Huang, Y.L. and Huang, T.F. (1990) Vasorelaxation of rat aorta by fargesone B isolated from the flower buds of *Magnolia fargesii*. *Asia Pacific J. Pharmacol.*, 5, 213–218.

Tomita, M. and Kazuka, M. (1967) Alkaloids of magnoliaceous plants. XXXVIII. Alkaloids of *Magnolia grandiflora*. 4. *Yakugaku Zasshi*, 87, 1134–1137.

Tomita, M. and Kugo, T. (1954) Alkaloids of magnoliaceous plants XI. Structure of magnolamine. *Pharm. Bull. (Japan)*, 2, 115–118.

Tomita, M. and Nakano, T. (1952a) Alkaloids of Manoliaceous plants III. Alkaloids of *Magnolia salicifolia*. *J. Pharm. Soc. Jpn.*, 72, 197–203.

Tomita, M. and Nakano, T. (1952b) Alkaloids of Manoliaceous plants V. Alkaloids of *Magnolia kobus*. *J. Pharm. Soc. Jpn.*, 72, 727–731.

Tomita, M. and Nakano, T. (1952c) Alkaloids of Manoliaceous plants VI. Alkaloids of *Magnolia stellata*. *J. Pharm. Soc. Jpn.*, 72, 766–767.

Tomita, M. and Nakano, T. (1952d) Alkaloids of *Magnolia denudata*. *J. Pharm. Soc. Jpn.*, 72, 1260–1262.

Tomita, M. and Nakano, T. (1957) Alkaloids of the Magnolia family. *Planta Med.*, 5, 33–43.

Tomita, M., Inubushi, Y. and Yamagata, M. (1951) Alkaloids of Magnoliaceae. I. Alkaloids of *Magnolia obovata*. *J. Pharm. Soc. Jpn.*, 71, 1069–1075.

Tomita, M., Watanabe, Y. and Furukawa, H. (1961) Alkaloids of magnoliaceous plants XXV. Alkaloids of *Magnolia grandiflora* var. *laceolata*. *Yakugaku Zasshi*, 81, 144–146.

Tomita, M., Lu, S.-T., Wang, S.-J., Lee, C.-H. and Shih, H.-T. (1968) Alkaloids of magnoliaceous plants. XXXIX. Alkaloids of *Magnolia kachirachirai*. 3. *Yakugaku Zasshi*, 88, 1143–1147.

Tsuruga, T., Ebizuka, Y., Nakajima, J., Chun, Y.T., Noguchi, H., Iitaka, Y. and Sankawa, U. (1991) Biologically active constituents of *Magnolia salicifolia* inhibitors of induced histamine release from rat mast cells. *Chem. Pharm. Bull.*, 39, 3265–3271.

Watanabe, H., Ikeda, M., Watanabe, K. and Kikuchi, T. (1981) Effects on central dopaminergic systems of D-colaurine and D-reticuline, extracted from *Magnolia salicifolia*. *Planta Med.*, 42, 213–222.

Weevers, T.H. (1930) Quercetin in Magnoliaceae and its distribution in the vegetable kingdom. *Proc. Acad. Sci. Amsterdam*, 33, 778–785.

Wiedhopf, R.M., Young, M., Bianchi, E. and Cole, J.R. (1973) Tumor inhibiting agent from *Magnolia grandiflora* (*Magnoliaceae*) I. Parthenolide. *J. Pharm. Sci.*, 62, 345.

Yahara, S., Nishiyori, T., Kohda, A., Nohara, T. and Nishioka, I. (1991) Isolation and characterisation of phenolic compounds from Magnoliae cortex produced in China. *Chem. Pharm. Bull.*, 39, 2024–2036.

Yan, W.-M. (1979) Chemical components in the bark of *Magnolia rostrata* W.W. Smith. *Chin. Wu Hsueh Pao*, 21, 54–56.

Yang, M.-H., Blunden, G., Patel, A.V., Oneill, M.L. and Lewis, J.A. (1994) Coumarins and sesquiterpene lactones from *Magnolia grandiflora* leaves. *Planta Med.*, 60, 390.

Yang, T.-H. (1971) Alkaloid from *Magnolia coco* (Magnoliaceae). *J. Chin. Chem. Soc. (Taipei)*, 18, 91.

Yang, T.-H., and Liu, S.-C. (1973) Alkaloids of *Magnolia coco*. II. *Pei. I. Hsueh Pao*, 3, 121–125.

Yang, T.-H. and Lu, S.-T. (1963) Alkaloids of magnoliaceous plants XXXV. Alkaloids of *Magnolia kachirachirai 2.* Isolation of D-(+)-N-norarmepavine. *Yakugaku Zasshi*, 83, 22–25.

Yang, T.-H., Lu, S.-T. and Hisao, C.-Y. (1962) Alkaloids of magnoliaceous plants XXXII. Alkaloids of *Magnolia coco* and *Magnolia kachirachirai*, *Ibid.*, 816–820.

Yu, H.-J., Chen, C.-C. and Shieh, B.-J. (1998a) Two new constituents from the leaves of *Magnolia coco. J. Nat. Prod.*, 61, 1017–1019.

Yu, H.-J., Chen, C.-C. and Shieh, B.-J. (1998b) The constituents from the leaves of *Magnolia coco. J. Chinese Chem. Soc.*, 45, 773–778.

Yu, S.M., Chen, C.C., Huang, Y.L., Tsai, C.W., Lin, C.H., Huang, T.F. and Teng, C.M. (1990) Vasorelaxing effect in rat thoracic aorta caused by denudatin B, isolated from the Chinese herb, *Magnolia fargesii. Eur. J. Pharmacol.*, 187, 39–47.

Ziyaev, R., Abdusamatov, A. and Yunusov, S. (1975) Alkaloids of *Magnolia soulangeana. Khim. Prir. Soedin.*, 11, 528–529.

Ziyaev, R., Abdusamatov, A. and Ikromov, K. (1995) Alkaloids of *Magnolia kobus. Khim. Prir. Soedin.*, 2, 327–328.

Ziyaev, R., Shtonda, N.I., Sturua, M.D., Abdusamatov, A. and Tsakadze, D.M. (1999) Alkaloids of some *Magnolia* species. *Chem. Nat. Compounds*, 35, 366–367.

4 Bioactivity and Pharmacological Aspects of *Magnolia*

Yuji Maruyama, Yasushi Ikarashi and Hisashi Kuribara

Section I: Classification of the biological activities of Magnoliaceae species
    4.1.1   Introduction
    4.1.2   Biochemical pharmacology
    4.1.3   Cardiovascular pharmacology
    4.1.4   CNS pharmacology
    4.1.5   Renal pharmacology
    4.1.6   Gastrointestinal pharmacology
    4.1.7   Immunopharmacology and inflammation
    4.1.8   Antimicrobial activities
    4.1.9   Conclusion
Section II: Anti-asthmatic, anxiolytic and anti-ulcer effects of Saiboku-to and biogenic histamine
    4.2.1   Introduction
    4.2.2   Effects of Saiboku-to
    4.2.3   Conclusion
Section III: Behavioral determination of anxiolytic effect of herbal medicines as being caused by the prescribed *Magnolia* component
    4.3.1   Introduction
    4.3.2   Anxiolytic effect of Hange-koboku-to and Saiboku-to
    4.3.3   Conclusion
Section IV: Pharmacological characteristics of *Magnolia* extracts: magnolol and honokiol
    4.4.1   Introduction
    4.4.2   Central depressive effects of *Magnolia* extracts
    4.4.3   Central depressive effects of magnolol and honokiol
    4.4.4   Mechanisms of the central depressive effects of magnolol and honokiol
    4.4.5   Interaction with centrally acting drugs
    4.4.6   Benzodiazepine-like side effects of *Magnolia*-prescribed Kampo medicines
    4.4.7   Benzodiazepine-like side effects of honokiol
    4.4.8   Side effects of DHH-B
    4.4.9   Pharmacokinetics of magnolol and honokiol
    4.4.10 Metabolism of magnolol and honokiol
    4.4.11 Conclusion

SECTION I   CLASSIFICATION OF THE BIOLOGICAL ACTIVITIES OF *MAGNOLIACEAE* SPECIES

Yuji Maruyama

4.1.1   Introduction

Almost 70 studies describing specific bioactivities of extracts of magnoliaceous plants and their ingredients have been published over the last 25 years. Different species of Magnoliaceae, such as *Magnolia officinalis*, *M. obovata*, *M. sieboldii*, *M. grandiflora*, *M. coco*, *M. fargesii*, *M. flos* and *M. salicifolia* were investigated. In this section, 58 representative reports have been selected and classified into seven major categories on the basis of the pharmacological features described: (i) biochemical, (ii) cardiovascular, (iii) CNS, (iv) renal, (v) gastrointestinal, (vi) immunopharmacology and inflammatory pharmacology, and (vii) antimicrobial activities. The names of each extract and/or major components derived from the specific Magnoliaceae species are emphasized in bold type, and each study is briefly summarized.

4.1.2   Biochemical pharmacology

4.1.2.1   Sepsis: lipid peroxidation

To determine whether **magnolol** can modulate the course of sepsis, survival rate and biochemical parameters were analyzed in rats with sepsis using various treatment protocols. The intensity of lipid peroxidation in plasma, liver, and lung of septic rats was also attenuated in a treatment-dependent manner. Significant results of the study suggested that magnolol may be useful in the treatment of sepsis (Kong *et al.*, 2000).

4.1.2.2   Magnolol-induced {Ca^{2+}}$_i$ elevation

Wang and Chen (1998) explored the role of inositol trisphosphate in the signaling pathway that leads to the elevation of cytosolic-free Ca^{2+} in rat neutrophils stimulated with **magnolol**. They suggested that a pertussis toxin-insensitive inositol trisphosphate signaling pathway is involved in the magnolol-induced [Ca^{2+}]$_i$ elevation in rat neutrophils.

4.1.2.3   Cholesterol acyltransferase inhibitor

New types of cholesterol acyltransferase (ACAT) inhibitors were isolated from the extract of *Magnolia obovata* leaves, and identified as **obovatol**, **honokiol** and **magnolol**. The active compounds inhibit rat liver ACAT with IC$_{50}$ values of 42, 71, and 86 μM, respectively (Kwon *et al.*, 1997).

4.1.2.4   Prednisolone metabolism inhibitor

To identify the inhibitor of prednisolone metabolism contained in Saiboku-to, an Oriental herbal medicine, *in vitro* experiments with 11β-hydroxysteroid dehydrogenase (11β-HSD) were conducted using rat liver homogenate and cortisol as a typical substrate

(Homma *et al.*, 1994). The results suggested that **magnolol** might contribute to the inhibitory effects of Saiboku-to on prednisolone metabolism through inhibition of 11β-HSD.

4.1.2.5   *Costunolide: nitric-oxide synthase inhibitor*

Syringin 4-*O*-β-glucoside, a new phenylpropanoid glycoside, and **costunolide** isolated from the bark of *Magnolia sieboldii* exhibited strong nitric-oxide synthase inhibitory activity in the endotoxin-activated murine macrophage J774.1 (Park *et al.*, 1996).

4.1.2.6   *Antioxidant effect (1)*

Oxygen consumption and malondialdehyde production were measured to quantitate lipid peroxidation. The antioxidant effects of **magnolol** and **honokiol** purified from *Magnolia officinalis* were 1000 times higher than those of α-tocopherol. Both magnolol and honokiol also exhibited free radical scavenging activities in the diphenyl-*p*-picrylhydrazyl assay, but they were less potent than those of α-tocopherol (Lo *et al.*, 1994).

4.1.2.7   *Antioxidant effect (2)*

Major antioxidants, tetrahydrofurofuran lignans, in the fruits of *Forsythia suspensa* and the stem of *Magnolia coco*, were identified using the inhibition of Cu^{2+}-induced LDL oxidation as marker. Of these bioactive lignans, **pinoresinol**, **phillygenin** and **syringaresinol** were found to be more potent than **probucol**. **Sesamin** and **fargesin**, which do not contain phenol groups, were found to be much less active (Chen *et al.*, 1999).

4.1.2.8   *Antioxidant effect (3)*

The protective effect of **magnolol** on lipid peroxidation-suppressed sperm motility was examined. Magnolol significantly inhibited the generation of malondialdehyde, the end product of lipid peroxidation, in sperm (Lin *et al.*, 1995).

4.1.2.9   *Antioxidant effect (4)*

Microsomal lipid peroxidation induced by Fe(III)–ADP/NADPH and mitochondrial lipid peroxidation induced by Fe(III)–ADP/NADH were inhibited by **honokiol** and **magnolol**, neolignans in *Magnolia obovata*. The antioxidative activity of honokiol was more potent than that of magnolol (Haraguchi *et al.*, 1997).

4.1.2.10   *Inhibitors of skin tumor promotion*

Three neolignans, **magnolol**, **honokiol** and the new **monoterpenylmagnolol** exhibited inhibiting activities of Epstein–Barr virus early antigen activation induced by 12-*O*-tetradecanoylphorbol 13-acetate. The methanolic extract of *Magnolia officinalis* and magnolol exhibited remarkable inhibitory effects on mouse skin tumor promotion in an *in vivo* two-stage carcinogenesis test, suggesting that these samples might be valuable antitumor promoters (Konoshima *et al.*, 1991).

4.1.2.11   HPLC for Saiboku-to urine

Systematic analysis of the components in post-administrative Saiboku-to urine was conducted using high-performance liquid chromatography to elucidate some of the less well known effects of this traditional medicine. **Magnolol** and various other compounds were detected and identified as a result of this investigation (Homma *et al.*, 1997).

4.1.2.12   Rectal administration of Magnolia bark

To examine the level of **magnolol** in rabbit blood, *Magnolia* bark was administered rectally and results were compared with other routes of administration (Tan *et al.*, 1995).

4.1.2.13   Toxic component in Magnolia grandiflora

The toxic component in alcoholic extract of the wood of *Magnolia grandiflora* was identified as **menisperine** (chakranine and isocorydinium cation), a phenolic quarternary alkaloid (Rao and Davis, 1982).

4.1.2.14   TNF-α

Three TNF α-inhibitory lignans, **eudesmin, magnolin** and **lirioresinol-B dimethyl ether,** were isolated and identified from the flower buds of *Magnolia fargesii*. Inhibitory effects were determined on the basis of effects on TNF-α production in the LPS-stimulated murine macrophage cell line RAW264.7, and eudesmin showed the strongest activity (IC_{50} = 51 μM) (Chae *et al.*, 1998).

4.1.2.15   Anti-complement activity: tiliroside

Anti-complementary active components were isolated from *Magnolia fargesii* and identified as **tiliroside** exhibiting very potent activity of IC_{50} = 5.4 × 10^{-5} M when compared with a well-known inhibitor, rosmarinic acid, on the classical pathway of the complement system (Jung *et al.*, 1998a).

4.1.2.16   CINC-1 inhibition

An inhibitor in CINC-1 (cytokine-induced neutrophil chemoattractant-1) induction in LPS-stimulated rat kidney epithelioid NRK-52E cells was purified from the roots of *Sassurea lappa Clarke*, a herbal medicine used in Korean traditional prescriptions. This inhibitor was identified as **reynosin**, which was previously isolated and characterized from *Ambrosia confertiflora* DC and *Magnolia grandiflora* L (Jung *et al.*, 1998b).

4.1.2.17   Cold stress-averting activity

Shizandra ligands derived from *Magnolia* vine were examined to ascertain cell manifestations of disadaptation in animals exposed to chronic cold stress of 1.5-hour cooling to 28–30 °C for 28 days. The effect was evaluated by determining the level of pathological mitoses (PM). Shizandra administration inhibited stress-induced elevation of the PM level in the corneal epithelium and also promoted the normalization of DNA synthesis and mitotic activity in the tissues under examination (Mel'nik *et al.*, 1984).

4.1.2.18 *Catecholamine secretion*

The molecular mechanism of **honokiol**, extracted from the bark of *Magnolia obovata*, was studied using bovine adrenal chromaffin cells as a model system. Honokiol inhibits catecholamine secretion induced by carbachol and DMPP and to a lesser extent that induced by exposure to high K^+ and Ba^{2+}. The results suggest that honokiol interferes with the interaction between the acetylcholine receptor and its agonists and that honokiol may also affect the steps in exocytosis after intracellular calcium level has been raised (Liu *et al.*, 1989).

4.1.3 Cardiovascular pharmacology

4.1.3.1 *Anti-platelet activity (1)*

Two new lignans having platelet-activating-factor (PAF) antagonists were isolated from the flower buds of *Magnolia fargesii* and identified as **magnones A** and **B**. Through [^{3}H]PAF receptor binding assay, activities of these lignans were determined as IC_{50} values of 3.8×10^{-5} M and 2.7×10^{-5} M, respectively (Jung *et al.*, 1998c).

4.1.3.2 *Anti-platelet activity (2)*

The vasorelaxing agent **denudatin B**, an isomer of kadsurenone, inhibited the aggregation and ATP release of washed rabbit platelets caused by PAF in a concentration-dependent manner. IC_{50} on PAF (2 ng/ml)-induced aggregation was about 10 μg/ml. The experimental results indicate that at high concentration denudatin B exibits nonspecific anti-platelet action by inhibiting phosphoinositide breakdown induced by collagen and thrombin (Teng *et al.*, 1990a).

4.1.3.3 *Anti-platelet activity (3)*

Magnolol isolated from *Magnolia officinalis* exhibited anti-platelet activity (Teng *et al.*, 1990b); it inhibited norepinephrine-induced phasic and tonic contractions in thoracic aorta of rats. It was concluded that magnolol relaxes vascular smooth muscle by releasing endothelium-derived relaxing factor and by inhibiting calcium influx through voltage-gated calcium channels.

4.1.3.4 *Anti-platelet activity (4)*

Antiplatelet activity of **magnolol** and **honokiol** isolated from *Magnolia officinalis* was examined (Teng *et al.*, 1988). Both inhibited the aggregation and ATP release of rabbit platelet-rich plasma induced by collagen and arachidonic acid without affecting those induced by ADP, PAF or thrombin. The antiplatelet effect of these compounds was found to be due to an inhibitory effect on thromboxane formation and also an inhibition of intracellular calcium mobilization.

4.1.3.5   *Vasorelaxing agent*

Denudatin B, an antiplatelet agent, was isolated from the flower buds of *Magnolia fargesii*. The effects of this compound on the vasoconstriction of rat thoracic aorta induced by high potassium solution, norepinephrine and caffeine and its mode of action were examined. The results indicate that denudatin B relaxed vascular smooth muscle by inhibiting Ca^{2+} influx through voltage-gated and receptor-operated Ca^{2+} channels; its effect in increasing cGMP may enhance vasorelaxation (Yu *et al.*, 1990).

4.1.3.6   *Inotropic activity*

Inotropic activities of **(+/−)-higenamine** (Hig), **(+)-(*R*)-coclaurine** (Coc) and **(+)-(*S*)-reticuline** (Ret) were examined on isolated guinea-pig papillary muscle. Hig is a cardiotonic principle from aconite root, and Coc and Ret were derived from the dried buds of *Magnolia salicifolia* MAXIM. All these alkaloids possess a common chemical structure of tetrahydroisoquinoline. Coc and Ret demonstrated actions contrary to those of Hig as the Ca^{2+} curve clearly reveals (Kimura *et al.*, 1989).

4.1.3.7   *Myocardial protective effect*

The *in vivo* anti-arrhythmic and anti-ischemic effects of **honokiol**, an active component of *Magnolia officinalis*, were examined in coronary-ligated rats. The results suggested that honokiol may protect the myocardium against ischemic injury and suppress ventricular arrhythmia during ischemia and reperfusion (Tsai *et al.*, 1996). Later Hong *et al.* (1996) also found a similar effect of **magnolol** using the same procedure, and confirmed the conclusions reached by Tsai *et al.* (1996).

4.1.3.8   *Anti-angiogenic activities*

The effects of **magnosalin** (MSA) and **magnoshinin** (MSI), neolignans isolated from *Magnolia*, on tube formation of endothelial cells cultured in type I collagen gel during the angiogenic process were examined by Kobayashi *et al.* (1996). They found that MSA inhibited fetal bovine serum (FBS)-stimulated tube formation with a greater potency than MSI. The inhibitory effect of MSA on the action of FBS and the action of IL-1α were found to be different.

4.1.3.9   *Activity on the cardiovascular system*

Effects of the **aqueous extracts** of flowers and leaves of *Magnolia grandiflora* L. on the cardiovascular system were examined using different animal models. The study was prompted by the wide use of this plant in traditional Mexican medicine (Mellado *et al.*, 1980).

4.1.4   CNS pharmacology

4.1.4.1   *Anxiolytic activity*

Honokiol, a neolignan from *Magnolia* bark, exhibited central depressant action and, at much lower doses, anxiolytic activity (Kuribara *et al.*, 1998). The behavioral effects of honokiol on the characteristics were investigated using an improved plus-maze test. The results suggested that, in contrast to diazepam, honokiol possesses a selective anxiolytic effect without significantly eliciting motor dysfunction and sedation or disinhibition. Thus, honokiol exhibits an action partially different from that of diazepam.

The author's group had previously confirmed that anxiolytic activity of Saiboku-to, an oriental herbal medicine, was mainly due to honokiol derived from *Magnolia* (Kuribara *et al.*, 1999a). The goal of this study was to compare the anxiolytic potential of **honokiol** and **water extracts** of three different *Magnolia* samples; two *M. officinalis* and one *M. obovata*. The results suggested that honokiol is the chemical responsible for the anxiolytic effect of the water extract of *Magnolia* and that other chemicals, including magnolol in *Magnolia*, scarcely influence the effect of honokiol.

4.1.4.2 Honokiol (no diazepam-like side effects)

Use of the elevated plus-maze experiment and activity and traction tests in mice has revealed that seven daily treatments with 0.2 mg/kg and higher doses of **honokiol** produced an anxiolytic effect without causing a change in motor activity or muscle tone. This study sought to determine whether honokiol had diazepam-like side effects. The results suggested that honokiol is less likely than diazepam to induce physical dependence, central depression and amnesia at doses eliciting the anxiolytic effect. It was also speculated that honokiol may have no therapeutic effect in the treatment of convulsions (Kuribara *et al.*, 1999b).

4.1.4.3 Confirmation of the anxiolytic ingredient in Magnolia

An improved elevated plus-maze test in mice revealed that seven daily treatments with Hange-koboku-to (composed of extracts of five plants) and Saiboku-to (composed of extracts of 10 plants), both traditional Chinese medicines (called Kampo medicine in Japan), produced an anxiolytic activity. On the other hand, *Magnolia*-free prescriptions of the above two herbal medicines did not cause the anxiolytc activity. Thus, the effect was mainly due to the presence of **honokiol** derived from the two different *Magnolia* species prescribed (*M. officinalis* and *M. obovata*). The results confirmed that honokiol derived from *Magnolia* is the causal ingredient of the anxiolytic effect of the two Kampo medicines used in this study (Kuribara *et al.*, 2000).

4.1.4.4 Identification of anxiolytic agents in Saiboku-to

The principal active anxiolytic components in Saiboku-to have been isolated and identified as **magnolol** and **honokiol**. An improved plus-maze test showed that honokiol was at least 5000 times more potent than Saiboku-to when orally administered to mice for 7 days (Maruyama *et al.*, 1998).

4.1.4.5 CNS-acting muscle relaxant

Effects of **diphenyl** and **monophenyl compounds** on grip strength in mice and spinal reflexes in young chicks were investigated to determine the relationship between the structures and onset of the activities. **Magnolol** and **honokiol** extracted from *Magnolia officinalis* Thunb. were used. The results suggested that the presence of a hydroxyl group accelerates the onset and shortens the duration of muscle relaxant activity of diphenyl, whereas the presence of an allyl group shows the opposite effect (Watanabe *et al.*, 1983b,c, 1975, 1973).

4.1.4.6   *Hypoxia protection*

The protective effect of **magnolol**, a component of *M. officinalis*, on hypoxia-induced cell injury in cortical neuron–astrocyte mixed cultures was examined (Lee *et al.*, 1998). Magnolol, at doses of 10 and 100 μM, significantly reduced the KCN-induced LDH release in a dose-dependent manner, demonstrating the protective activity of magnolol for neurons against chemical hypoxic damage or necrotic cell death under the conditions described.

4.1.4.7   *CNS 5HT activity*

To examine the possibility that the depressant effect of **magnolol** may be exhibited through modulation of central serotonergic activity, the effect of this compound on 35 mM K^+-stimulated $[H^3]5HT$ release from rat hippocampal and frontal cortical slices was studied. It was concluded that the suppression of brain 5HT release by magnolol is site-specific, and the suppression of cortical 5HT release by magnolol is not accomplished via 5HT autoreceptors at 5HT terminals (Tsai *et al.*, 1995).

4.1.4.8   *CNS dopaminergic effect*

An intracerebroventricular injection of **d-coclaurine** extracted from *Magnolia salicifolia* produced a slight increase in 3,4-dihdroxyphenylacetic acid and a significant increase in the homovanillic acid level in the mouse striatum. Another alkaloid, **d-reticuline** increased only the homovanillic acid level. The results obtained through metabolic examinations suggested that *d*-coclaurine blocks postsynaptic but not presynaptic dopamine receptors in the mouse striatum (Watanabe *et al.*, 1981, 1983a).

4.1.4.9   *Anticonvulsive activity*

Antispasmodic activities of ethyl ether and **hydroalcoholic extracts of Magnolia grandiflora L. seeds** were studied in adult male Wistar rats. The results exhibited abolition of the extensor reflex in the maximal electrically induced seizure test in 50% and 40% of the experimental animals. They significantly prolonged the sleeping time induced by pentobarbital and only the ethanol extract induced hypothermia. These results suggest that the chemical constituents of this plant could be useful for treatment with epileptic patients during convulsive seizures (Bastidas Ramirez *et al.*, 1998).

4.1.4.10   *GABA$_A$ receptor*

The effects of **magnolol** and **honokiol** on $[^3H]$muscimol and $[^3H]$flunitrazepam binding sites were examined using forebrain and cerebellar membrane preparations in the rat (Squires *et al.*, 1999). The results suggested that potentiation of GABAergic neurotransmission by these two compounds is probably responsible for their previously reported anxiolytic and central depressant effects.

4.1.4.11 Effect of decreasing body temperature

The effects of **magnolol**, isolated and purified from the cortex of *M. officinalis* Rehd. et Wils, on thermoregulation and hypothalamic release of 5HT were assessed by *in vivo* microdialysis in normothermic rats and in febrile rats treated with interleukin-1β (Hsieh *et al.*, 1998). The data suggest that magnolol decreases body temperature (due to increased heat loss and decreased heat production) by reducing 5HT release in rat hypothalamus.

4.1.5 Renal pharmacology

4.1.5.1 Belgian (Chinese herb) nephropathy

Several patients with renal failure were admitted to Brussels hospitals over recent years. The progressive interstitial fibrosis with tubular atrophy seen in these patients has been ascribed to the slimming therapy preceding the pathology. The progress and extent of nephropathy were extreme. The ingestion of *Aristolochia fangchi* instead of the prescribed *Stephania tetrandra*, one of the components of the slimming therapy, was advanced in literature as a hypothesis for the etiology of the nephropathies. Medicinal plants such as those suggested contain secondary metabolites (bis)-benzylisoquinoline alkaloids, dihydroxydiallyl biphenyls, and aristolochic acids with strong pharmacological (and possibly toxic) actions. The authors express concern that prohibiting (temporarily) the use of three Chinese herbs (*Stephania tetrandara, Aristolochia fangchi* and *Magnolia officinalis*) is not adequate for safety in responsible common health care (Violon, 1997; Schmeiser *et al.*, 1996; Vanherweghem, 1994; Vanherweghem *et al.*, 1993).

4.1.6 Gastrointestinal pharmacology

4.1.6.1 Anti-emetic action

Magnolol and **honokiol**, biphenyl compounds, were isolated as anti-emetic principles from the methanolic extract of *Magnolia obovata* bark as well as shogaols and gingerols from *Zingiber officinale* rhizome. Some phenylpropanoids with allyl side chains were found to show the same activity. They inhibited the emetic action induced by the oral administration of copper sulfate pentahydrate to leopards and ranid frogs (Kawai *et al.*, 1994).

4.1.7 Immunopharmacology and inflammation

4.1.7.1 Inhibitory effect on neutrophil adhesion

Magnolol has been shown to protect rat heart from ischemia/reperfusion injury (Shen *et al.*, 1998). Neutrophil adhesion plays a critical role during this inflammatory response. The author's group proposed that the inhibitory effect of magnolol on neutrophil adhesion to the extracellular matrix is mediated, at least in part, by inhibition of the accumulation of reactive oxygen species, which in turn suppresses the upregulation of Mac-1 which is essential for neutrophil adhesion.

4.1.7.2 Anti-inflammatory effect

A23187-induced pleurisy in mice was used to investigate the anti-inflammatory effect of **magnolol**, isolated from Chinese medicine Hou p'u (cortex of *Magnolia officinalis*). A23187-induced protein leakage was reduced by magnolol (10 mg/kg, i.p.), indomethacin (10 mg/kg, i.p.) and BW755C (30 mg/kg, i.p.). The inhibitory effect is proposed to be, at least partly, dependent on the reduction of the formation of eicosanoid mediators at the inflammatory site (Wang *et al.*, 1992, 1995).

4.1.7.3 Anti-allergic activity

Extracts of the flower buds of M. *salicifolia* were used for identification of anti-allergic compounds. The extracts exhibited potent anti-allergy effects in a passive cutaneous anaphylaxis test. The bioactive constituents of this medicinal drug were isolated by monitoring their activities with an *in vitro* bioassay system in which an inhibitory effect on histamine release was induced by compound 48/80 or Com. Of the ten isolated compounds, **magnosalicin** was found to be a new neolignan (Tsuruga *et al.*, 1991).

4.1.8 Antimicrobial activities

Three phenolic constituents of *M. grandiflora* L. were shown to exhibit significant antimicrobial activity using an agar well diffusion assay. **Magnolol, honokol, and 3,5'-diallyl-2'-hydroxy-4-methoxybiphenyl** exhibited significant activity against Gram-positive and acid-fast bacteria and fungi (Clark *et al.*, 1981).

4.1.8.1 Anti-Helicobacter pylori activity

The extracts of *Coptidis japonica* (rhizoma), *Eugenia cryophyllata* (flower), *Rheum palmatum* (rhizoma), ***Magnolia officinalis*** (cortex) and *Rhus javanica* (galla rhois) potently inhibited the growth of *Helicobacter pylori* (HP). However, with the exception of Galla rhois, these herbal extracts showed no inhibitory effect on HP urease. Minimum inhibitory concentrations of decursin and decursinol angelate were found to be 6–20 µg/ml (Bae *et al.*, 1998; Zhang *et al.*, 1992).

4.1.8.2 Periodontopathic microorganisms

Magnolol and **honokiol** from the stem bark of *M. obovata* were assessed for their antimicrobial activity against periodontopathic microorganisms, *Porphyromonas gingivalis*, *Prevotella gingivalis*, *Actinobacillus actinomycetemcomitans*, *Capnocytophaga gingivalis*, and *Veillonella disper*, and for cytotoxicity against human gingival fibroblasts and epithelial cells. It has been suggested that both these compounds may have potential therapeutic use as safe oral antiseptics for the prevention and the treatment of peridontal diseases (Chang *et al.*, 1998).

4.1.8.3 Antimicrobial activities

Three phenolic constituents of *Magnolia grandiflora* L. were shown to exhibit significant antimicrobial activity using an agar well diffusion assay. **Magnolol, honokiol, and**

3,5'-diallyl-2'-hydroxy-4-methoxybiphenyl exhibited significant activity against Gram-positive and acid-fast bacteria and fungi (Clark *et al.*, 1981).

4.1.9 Conclusion

It was possible to categorize the 58 selected studies into seven major categories on the basis of the pharmacological actions investigated. Of these reports, more than 80% were found to belong to three categories of features: biochemical (37.5%), cardiovascular (20.8%) and CNS pharmacology (25.0%). Almost 90% of these properties were derived from magnolol (54.2%) and honokiol (33.3%), and other identified compounds and extracts (12.5%). Specific properties of honokiol were introduced in the recently published "Overview of the pharmacological features of honokiol" (Maruyama and Kuribara, 2000). The evidence presented in this and other studies points to a wide range of bioactivities characteristic of magnolol and honokiol. The problems associated with the onset of side effects were discussed in the Renal Pharmacology subsection, and relevant findings were introduced (Vanherweghem, 1994; Vanherweghem *et al.*, 1993; Schmeiser *et al.*, 1996; Violon, 1997). Although traditional herbal materials, including magnolol, and their ingredients have been utilized for thousands of years for treatment of a wide variety of clinical disorders, a non-placebo scientific assessment of the medicinal plants is necessary to determine their safety and application. An overview of literature on bioactivities of *Magnolia* plant extracts and their principal components presented in this section is the first step in this direction.

References for section I

Bae, E.A., Han, M.J., Kim, N.J. and Kim, D.H. (1998) Anti-*Helicobacter pylori* activity of herbal medicines. *Biol. Pharm. Bull.*, 21(9), 990–992.

Bastidas Ramirez, B.E., Navarro Ruiz, N., Quezada Arellano, J.D., Ruiz Madrigal, B., Villanueva Michel, M.T. and Garzon, P. (1998) Anticonvulsant effects of *Magnolia grandiflora* L. in the rat. *J. Ethnopharmacol.*, 61(2), 143–152.

Chae, S.H., Kim, P.S., Cho, J.Y., Park, J.S., Lee, J.H., Yoo, E.S., Baik, K.U., Lee, J.S. and Park, M.H. (1998) Isolation and identification of inhibitory compounds on TNF-α production from *Magnolia fargesii. Arch. Pharm. Res.*, 21(1), 67–69.

Chang, B., Lee, Y., Ku, Y., Bae, K. and Chung, C. (1998) Antimicrobial activity of magnolol and honokiol against periodontopathic microorganisms. *Planta Med.*, 64(4), 367–369.

Chen, C.C., Chen, H.Y., Shiao, M.S., Lin, Y.L., Kuo, Y.H. and Ou, J.C. (1999) Inhibition of low density lipoprotein oxidation by tetrahydrofurofuran lignans from *Forsythia suspensa* and *Magnolia coco. Planta Med.*, 65(8), 709–711.

Clark, A.M., El-Feraly, F.S. and Li, W.S. (1981) Antimicrobial activity of phenolic constituents of *Magnolia grandiflora* L. *J. Pharm. Sci.*, 70(8), 951–952.

Haraguchi, H., Ishikawa, H., Shirataki, N. and Fukuda, A. (1997) Antiperoxidative activity of neolignans from *Magnolia obovata. J. Pharm. Pharmacol.*, 49(2), 209–212.

Homma, M., Oka, K., Niitsuma, T. and Itoh, H. (1994) A novel 11β-hydroxysteroid dehydrogenase inhibitor contained in saiboku-to, a herbal remedy for steroid-dependent bronchial asthma. *J. Pharm. Pharmacol.*, 46(4), 305–309.

Homma, M., Oka, K., Taniguchi, C., Niitsuma, T. and Hayashi, T. (1997) Systematic analysis of post-administrative saiboku-to urine by liquid chromatography to determine pharmacokinetics of traditional Chinese medicine. *Biomed. Chromatogr.*, 11(3), 125–131.

Hong, C.Y., Huang, S.S. and Tsai, S.K. (1996) Magnolol reduces infarct size and suppresses ventricular arrhythmia in rats subjected to coronary ligation. *Clin. Exp. Pharmacol. Physiol.*, 23(8), 660–664.

Hsieh, M.T., Chueh, F.Y. and Lin, M.T. (1998) Magnolol decreases body temperature by reducing 5-hydroxytryptamine release in the rat hypothalamus. *Clin. Exp. Pharmacol. Physiol.*, 25(10), 813–817.

Jung, K.Y., Oh, S.R., Park, S.H., Lee, I.S., Ahn, K.S., Lee, J.J. and Lee, H.K. (1998a) Anti-complement activity of tiliroside from the flower buds of *Magnolia fargesii. Biol. Pharm. Bull.*, 21(10), 1077–1078.

Jung, J.H., Ha, J.Y., Min, K.R., Shibata, F., Nakagawa, H., Kang, S.S., Chang, I.M. and Kim, Y. (1998b) Reynosin from Sassurea lappa as inhibitor on CINC-1 induction in LPS-stimulated NRK-52E cells. *Planta Med.*, 64(5), 454–455.

Jung, K.Y., Kim, D.S., Oh, S.R., Park, S.H., Lee, I.S., Lee, J.J., Shin, D.H. and Lee, H.K. (1998c) Magnone A and B, novel anti-PAF tetrahydrofuran lignans from the flower buds of *Magnolia fargesii. J. Nat. Prod.*, 61(6), 808–811.

Kawai, T., Kinoshita, K., Koyama, K. and Takahashi, K. (1994) Anti-emetic principles of *Magnolia obovata* bark and *Zingiber officinale* rhizome. *Planta Med.*, 60(1), 17–20.

Kimura, I., Chui, L.H., Fujitani, K., Kikuchi, T. and Kimura, M. (1989) Inotropic effects of (+/−)-higenamine and its chemically related components, (+)-R-coclaurine and (+)-S-reticuline, contained in the traditional sino-Japanese medicines "bushi" and "shin-i" in isolated guinea pig papillary muscle. *Jpn. J. Pharmacol.*, 50(1), 75–78.

Kobayashi, S., Kimura, I. and Kimura, M. (1996) Inhibitory effect of magnosalin derived from *Flos magnoliae* on tube formation of rat vascular endothelial cells during the angiogenic process. *Biol. Pharm. Bull.*, 19(10), 1304–1306.

Kong, C.W., Tsai, K., Chin, J.H., Chan, W.L. and Hong, C.Y. (2000) Magnolol attenuates peroxidative damage and improves survival of rats with sepsis. *Shock*, 13(1), 24–28.

Konoshima, T., Kozuka, M., Tokuda, H., Nishino, H., Iwashima, A., Haruna, M., Ito, K. and Tanabe, M. (1991) Studies on inhibitors of skin tumor promotion, IX. Neolignans from *Magnolia officinalis. J. Nat. Prod.*, 54(3), 816–822.

Kuribara, H., Stavinoha, W.B. and Maruyama, Y. (1998) Behavioural pharmacological characteristics of honokiol, an anxiolytic agent present in extracts of *Magnolia* bark, evaluated by an elevated plus-maze test in mice. *J. Pharm. Pharmacol.*, 50(7), 819–826.

Kuribara, H., Kishi, E., Hattori, N., Yuzurihara, M. and Maruyama, Y. (1999a) Application of the elevated plus-maze test in mice for evaluation of the content of honokiol in water extracts of magnolia. *Phytother. Res.*, 13(7), 593–596.

Kuribara, H., Stavinoha, W.B. and Maruyama, Y. (1999b) Honokiol, a putative anxiolytic agent extracted from magnolia bark, has no diazepam-like side-effects in mice. *J. Pharm. Pharmacol.*, 51(1), 97–103.

Kuribara, H., Kishi, E., Hattori, N., Okada, M. and Maruyama, Y. (2000) The anxiolytic effect of two oriental herbal drugs in Japan attributed to honokiol from *Magnolia* bark. *J. Pharm. Pharmacol.*, 52, 1425–1429.

Kwon, B.M., Kim, M.K., Lee, S.H., Kim, J.A., Lee, I.R., Kim, Y.K. and Bok, S.H. (1997) Acyl-CoA: cholesterol acyltransferase inhibitors from *Magnolia obovata. Planta Med.*, 63(6), 550–551.

Lee, M.M., Hseih, M.T., Kuo, J.S., Yeh, F.T. and Huang, H.M. (1998) Magnolol protects cortical neuronal cells from chemical hypoxia in rats. *Neuroreport*, 9(15), 3451–3456.

Lin, M.H., Chao, H.T. and Hong, C.Y. (1995) Magnolol protects human sperm motility against lipid peroxidation: a sperm head fixation method. *Arch. Androl.*, 34(3), 151–156.

Liu, P.S., Chen, C.C. and Kao, L.S. (1989) Multiple effects of honokiol on catecholamine secretion from adrenal chromaffin cells. *Proc. Natl. Sci. Counc. Repub. China B*, 13(4), 307–313.

Lo, Y.C., Teng, C.M., Chen, C.F., Chen, C.C. and Hong, C.Y. (1994) Magnolol and honokiol isolated from *Magnolia officinalis* protect rat heart mitochondria against lipid peroxidation. *Biochem. Pharmacol.*, 47(3), 549–553.

Maruyama, Y. and Kuribara, H. (2000) Overview of the pharmacological features of honokiol. *CNS Drug Rev.*, 6(1), 35–44.

Maruyama, Y., Kuribara, H., Morita, M., Yuzurihara, M. and Weintraub, S.T. (1998) Identification of magnololl and honokiol in the anxiolytic extracts of Saiboku-to, an oriental herbal medicine. *J. Nat. Prod.*, 61, 135–138.

Mellado, V., Chavez Soto, M.A. and Lozoya, X. (1980) Pharmacological screening of the aqueous extracts of *Magnolia grandiflora* L. *Arch. Invest. Med. (Mex.)*, 11(3), 335–346.

Mel'nik, E.I., Timoshin, S.S. and Lupandin, A.V. (1984) Effect of magnolia vine lignans on cell division processes in the corneal and lingual epithelium of white rats subjected to long-term exposure to cold stress [in Russian]. *Biull. Eksp. Biol. Med.*, 98(12), 718–720.

Park, H.J., Jung, W.T., Basnet, P., Kadota, S. and Namba, T. (1996) Syringin 4-O-β-glucoside, a new phenylpropanoid glycoside, and costunolide, a nitric oxide synthase inhibitor, from the stem bark of *Magnolia sieboldii*. *J. Nat. Prod.*, 59(12), 1128–1130.

Rao, K.V. and Davis, T.L. (1982) Constitutents of *Magnolia grandiflora*. III. Toxic principle of the wood. *J. Nat. Prod.*, 45(3), 283–287.

Schmeiser, H.H., Bieler, C.A., Wiessler, M., van Ypersele de Strihou, C. and Cosyns, J.P. (1996) Detection of DNA adducts formed by aristolochic acid in renal tissue from patients with Chinese herbs nephropathy. *Cancer. Res.*, 56(9), 2025–2028.

Shen, Y.C., Sung, Y.J. and Chen, C.F. (1998) Magnolol inhibits Mac-1 (CD11b/CD18)-dependent neutrophil adhesion: relationship with its antioxidant effect. *Eur. J. Pharmacol.*, 343(1), 79–86.

Squires, R.F., Ai, J., Witt, M.R., Kahnberg, P., Saederup, E., Sterner, O. and Nielsen, M. (1999) Honokiol and magnolol increase the number of [3H] muscimol binding sites three-fold in rat forebrain membrances *in vitro* using a filtration assay, by allosterically increasing the affinities of low-affinity sites. *Neurochem. Res.*, 24(12), 1593–1602.

Tan, Y., Lu, W., Zhao, S., Hu, Y. and Ma, Z. (1995) Research on rectal administration of bark of official *Magnolia* [in Chinese]. *Chung Kuo Chung Yao Tsa Chih*, 20(1), 30–32, 62.

Teng, C.M., Chen, C.C., Ko, F.N., Lee, L.G., Huang, T.F., Chen, Y.P. and Hsu, H.Y. (1988) Two antiplatelet agents from *Magnolia officinalis*. *Thromb Res.*, 50(6), 757–765.

Teng, C.M., Yu, S.M., Chen, C.C., Huang, Y.L. and Huang, T.F. (1990a) Inhibition of thrombin- and collagen-induced phosphoinositides breakdown in rabbit platelets by a PAF antagonist—denudatin B, an isomer of kadsurenone. *Thromb Res.*, 59(1), 121–130.

Teng, C.M., Yu, S.M., Chen, C.C., Huang, Y.L. and Huang, T.F. (1990b) EDRF-release and Ca²⁺-channel blockade by magnolol, an antiplatelet agent isolated from Chinese herb *Magnolia officinalis*, in rat thoracic aorta. *Life Sci.*, 47(13), 1153–1161.

Tsai, T.H., Lee, T.F., Chen, C.F. and Wang, L.C. (1995) Modulatory effects of magnolol on potassium-stimulated 5-hydroxytryptamine release from rat cortical and hippocampal slices. *Neurosci Lett.*, 186(1), 49–52.

Tsai, S.K., Huang, S.S. and Hong, C.Y. (1996) Myocardial protective effect of honokiol: an active component in *Magnolia officinalis*. *Planta Med.*, 62(6), 503–506.

Tsuruga, T., Ebizuka, Y., Nakajima, J., Chun, Y.T., Noguchi, H., Iitaka, Y. and Sankawa, U. (1991) Biologically active constituents of *Magnolia salicifolia*: inhibitors of induced histamine release from rat mast cells. *Chem. Pharm. Bull. (Tokyo)*, 39(12), 3265–3271.

Vanherweghem, J.L. (1994) A new form of nephropathy secondary to the absorption of Chinese herbs [in French]. *Bull. Mem. Acad. R. Med. Belg.*, 149(1–2), 128–135, discussion 135–140.

Vanherweghem, J.L., Depierreux, M., Tielemans, C., *et al.* (1993) Rapidly progressive interstitial renal fibrosis in young women: association with slimming regimen including Chinese herbs. *Lancet*, 341(8842), 387–391.

Violon, C. (1997) Belgian (Chinese herb) nephropathy: why? *J. Pharm. Belg.*, 52(1), 7–27.

Wang, J.P. and Chen, C.C. (1998) Magnolol induces cytosolic-free Ca²⁺ elevation in rat neutrophils primarily via inositol trisphosphate signalling pathway. *Eur. J. Pharmacol.*, 352, 329–334.

Wang, J.P., Hsu, M.F., Raung, S.L., Chen, C.C. Kuo, J.S. and Teng, C.M. (1992) Anti-inflammatory and analgesic effects of magnolol. *Naunyn. Schmiedebergs Arch. Pharmacol.*, **346**(6), 707–712.

Wang, J.P., Ho, T.F., Chang, L.C. and Chen, C.C. (1995) Anti-inflammatory effect of magnolol, isolated from *Magnolia officinalis*, on A23187-induced pleurisy in mice. *J. Pharm. Pharmacol.*, **47**(10), 857–860.

Watanabe, H., Ikeda, M., Watanabe, K. and Kikuchi, T. (1981) Effects on central dopaminergic systems of *d*-coclaurine and *d*-reticuline, extracted from *Magnolia salicifolia*. *Planta Med.*, **42**(3), 213–222.

Watanabe, H., Watanabe, K. and Kikuchi, T. (1983a) Effects of *d*-coclaurine and *d*-reticuline, benzyltetrahydroisoquinoline alkaloids, on levels of 3,4-dihydroxy phenylacetic acid and homovanillic acid in the mouse striatum. *J. Pharmacobiodyn.*, **6**(10), 793–796.

Watanabe, H., Watanabe, K. and Hagino, K. (1983b) Chemostructural requirement for centrally acting muscle relaxant effect of magnolol and honokiol, neolignane derivatives. *J. Pharmacobiodyn.*, **6**(3), 184–190.

Watanabe, K., Watanabe, H., Goto, Y., Yamaguchi, M., Yamamoto, N. and Hagino, K. (1983c) Pharmacological properties of magnolol and honokiol extracted from *Magnolia officinalis*: central depressant effects. *Planta Med.*, **49**(2), 103–108.

Watanabe, K., Goto, Y. and Yoshitomi, K. (1973) Central depressant effects of the extracts of magnolia cortex. *Chem. Pharm. Bull. (Tokyo)*, **21**(8), 1700–1708.

Watanabe, K., Watanabe, H.Y., Goto, Y., Yamamoto, N. and Yoshizaki, M. (1975) Studies on the active principles of magnolia bark. Centrally acting muscle relaxant activity of magnolol and honokiol. *Jpn. J. Pharmacol.*, **25**(5), 605–607.

Yu, S.M., Chen, C.C., Huang, Y.L., Tsai, C.W., Lin, C.H., Huang, T.F. and Teng, C.M. (1990) Vasorelaxing effect in rat thoracic aorta caused by denudatin B, isolated from the Chinese herb, *Magnolia fargesii*. *Eur. J. Pharmacol.*, **187**(1), 39–47.

Zhang, L., Yang, L.W. and Yang, L.J. (1992) Relation between *Helicobacter pylori* and pathogenesis of chronic atrophic gastritis and the research of its prevention and treatment [in Chinese]. *Chung Kuo Chung Hsi I Chieh Ho Tsa Chih*, **12**(9), 521–523, 515–516.

SECTION II   ANTI-ASTHMATIC, ANXIOLYTIC AND ANTI-ULCER EFFECTS OF SAIBOKU-TO AND BIOGENIC HISTAMINE

Yasushi Ikarashi

4.2.1   Introduction

Saiboku-to is a traditional oriental herbal medicine containing Magnoliae Cortex as one of the constituent herbs (Table 4.1). It comprises the following ten dried medical herbs with composition ratios given in parentheses: Bupleuri Radix (20.59%, root of *Bupleurum falcatum* Linné); Pinelliae Tuber (14.71%, tuber of *Pinellia ternata* Breitenbach); Hoelen (14.71%, fungus of *Poria cocos* Wolf); Scutellariae Radix (8.82%, root of *Scutellaria baicalensis* Georgi); Magnoliae Cortex (8.82%, bark of *Magnolia obovata* Thunberg); Zizyphi Fructus (8.82%, fruit of *Zizyphus jujuba* Miller var. *inermis* Rehder); Ginseng Radix (8.82%, root of *Panax ginseng* C.A. Meyer); Glycyrrhizae Radix (5.88%, root of *Glycyrrhiza uralensis* Fisher); Perillae Herba (5.88%, herb of *Perilla frutescens* Britton var. *acuta* Kudo); and Zingiberis Rhizoma (2.94%, rhizome of *Zingiber officinale* Roscoe). This remedy has been utilized for the treatment of bronchial asthma and anxiety-related disorders. Recently, anti-asthmatic, anxiolytic

Table 4.1 Herbal composition of Saiboku-to

Herbal name	Plant name	Family	Part used	Composition	
				g	% (w/w)
Bupleuri Radix	*Bupleurum falcatum* Linné	Umbelliferae	Root	7.0	20.56
Pinelliae Tuber	*Pinellia ternata* Breitenbach	Araceae	Tuber	5.0	14.71
Hoelen	*Poria cocos* Wolf	Polyporaceae	Fungus	5.0	14.71
Scutellariae Radix	*Scutellaria baicalensis* Georgi	Labiatae	Root	3.0	8.82
Magnoliae Cortex	*Magnolia obovata* Thunberg	Magnoliaceae	Bark	3.0	8.82
Zizyphi Fructus	*Zizyphus jujuba* Miller var. *inermis* Rehder	Rhamnaceae	Fruit	3.0	8.82
Ginseng Radix	*Panax ginseng* C.A. Meyer	Araliaceae	Root	3.0	8.82
Glycyrrhizae Radix	*Glycyrrhiza uralensis* Fisher	Leguminosae	Root	2.0	5.88
Perillae Herba	*Perilla frutescens* Britton var. *acuta* Kudo	Labiatae	Herb	2.0	5.88
Zingiberis Rhizoma	*Zingiber officinale* Roscoe	Zingiberaceae	Rhizome	1.0	2.94

and anti-ulcer effects of this compound have been confirmed by scientific evidence. These effects of Saiboku-to seem to be related to the inhibition of effects induced by biogenic histamine.

In the present review, we describe recent findings of the effect of Saiboku-to on histamine-related functions including experimentally induced asthma, anxiety and ulcer. Finally, we introduce the effects of constituent herbs of Saiboku-to on histamine release from rat peritoneal mast cells, in order to elucidate the relative importance of the ten constituent herbs in the anti-histamine release effect of Saiboku-to.

4.2.2 Effects of Saiboku-to

4.2.2.1 Anti-asthmatic effect

Saiboku-to has been used clinically for the treatment of bronchial asthma, which is related to type I allergic reaction. Umesato (1984) and Nishiyori *et al.* (1985) have reported the inhibitory effects of Saiboku-to using various immunological reactions in laboratory animals: e.g. 48-h homologous passive cutaneous anaphylaxis (PCA); antigen-induced histamine release in the peritoneal cavities of rats passively sensitized with anti-dinitrophenylated ascaris extract (DNP-As)-IgE serum; 7-day homologous PCA in guinea-pigs mediated by anti-benzylpenicilloyl bovine γ-globulin (BPO-BGG-IgE) serum; experimentally caused asthma in guinea-pigs; Schultz-Dale reaction in guinea-pig tracheal muscle; and antigen-induced histamine release from sensitized guinea-pig lung. Some (Bupleuri Radix, Glycyrrhizae Radix, Scutellariae Radix and Zizyphi Fructus) of the constituent herbs in Saiboku-to have also been shown to inhibit 48-h PCA in rats (Koda *et al.*, 1982). However, the mechanisms of the inhibitory effects on these reactions have not been studied. Subsequent experimental results reported by Toda *et al.* (1988) support these studies as they indicate that Saiboku-to (10^{-4} to 10^{-2} g/ml) appears to be an effective inhibitor of histamine release and degranulation from mouse peritoneal mast cells induced by Compound 48/80. More recently, the effect of Saiboku-to on histamine release was directly investigated using IgE-sensitized rat peritoneal mast cells as a typical experimental model of

Table 4.2 Effect of Saiboku-to on antigen-induced histamine release from IgE-sensitized peritoneal mast cells in rats

Group	% Release	Inhibition rate (%)
Antigen (10 µg/ml DNP-BSA)	29.14 ± 1.55	0 (Control)
0.1 mg/ml Saiboku-to + antigen	13.79 ± 2.12**	53
0.3 mg/ml Saiboku-to + antigen	7.92 ± 2.26***	73
1.0 mg/ml Saiboku-to + antigen	2.98 ± 3.55***	90

Rats were injected intraperitoneally with 1.0 ml of monoclonal anti-dinitrophenyl (DNP) IgE (titer: 1000). Twenty-four hours later, IgE-sensitized peritoneal mast cells were collected for use in the study. 50 µl of Saiboku-to (final concentration: 0.1, 0.3 and 1.0 mg/ml) was added to 400 µl of the cell suspension adjusted to 1×10^5 cells/ml. After incubation at 37 °C for 10 min, histamine release from the mast cells was elicited by addition of antigen, 50 µl of DNP-BSA (10 µg/ml) to Saiboku-to-pretreated cell suspension. For details see Ikarashi *et al.* (2001a). The total histamine level in mast cells used in the present study was 1096.8 ± 115.1 ng. Histamine release was expressed as percentage of total histamine level. Basal histamine release was 5.88 ± 1.51% of the total histamine level. The inhibition effect of each concentration of Saiboku-to on antigen-induced histamine release was expressed as a percentage of the antigen-induced histamine release level. Data are expressed as mean ± SE ($n = 4$). Statistical significance was assayed by a one-way ANOVA, followed by Bonferroni multiple-comparison test: ** $p < 0.01$ and *** $p < 0.001$ compared to antigen-induced control group.

type I allergy reaction (Ikarashi *et al.*, 2001a). Saiboku-to dose-dependently inhibited the antigen-induced histamine release from the IgE-sensitized mast cells at 0.1–1.0 mg/ml (Table 4.2). These results suggest that the anti-allergic effect of Saiboku-to is, at least partially, due to inhibition of histamine release from the mast cells.

In recent years, bronchial asthma has come to be regarded pathologically as a chronic inflammatory disease of the respiratory tract. Inhalational steroids and anti-inflammatory drugs are recognized as being effective against bronchial asthma. In asthmatic guinea-pigs sensitized to ovalbumin, 7-day administration of Saiboku-to inhibits the asthmatic response to an antigen challenge, and the number of eosinophils in fluid obtained by bronchoalveolar lavage 4 h after antigen challenge is decreased while the infiltration of eosinophils and T lymphocytes into the lung parenchyma is inhibited (Tohda *et al.*, 1999). These findings suggest that Saiboku-to also has the potential to become a useful anti-inflammatory drug for treatment of bronchial asthma. Homma *et al.* (1993, 1994) suggest that magnolol in Magnoliae Cortex is an active component that acts as an inhibitor of 11β-hydroxysteroid dehydrogenase and T lymphocyte proliferation, resulting in corticosteroid sparing. Nakajima *et al.* (1993) have reported that Saiboku-to spared the downregulation of glucocorticoid receptor of human lymphocytes, plasma ACTH, and cortisol levels. These findings lend support to the use of this preparation in the treatment of corticosteroid-dependent severe asthma in Japan and former Czechoslovakia. The compound serves to reduce the maintenance doses of corticosteroid—the "steroid sparing" effect.

There is evidence that nitric oxide (NO) is a mediator of non-adrenergic non-cholinergic (NANC) stimulation-induced relaxations of the anococcygeus muscle in rats (Li and Rand, 1989) and mice (Gibson *et al.*, 1990). In the airway, NO is involved in NANC relaxation of tracheal smooth muscle in guinea-pigs (Tucker *et al.*, 1990; Li and Rand, 1991). The inhibitory effect of NO on cholinergic neurotransmission in rat trachea, a process involved in the induction of asthma, has also been suggested (Sekizawa *et al.*, 1993). A similar regulation of NO by acetylcholine (ACh) release is also found in N-methyl-D-aspartate-induced cholinergic neurotransmission in rat striatum (Ikarashi *et al.*, 1998). Ichinose *et al.* (1996) demonstrated that IgE enhances cholinergic bronchial contraction via facilitation of ACh release from cholinergic nerves in human airways *in vitro*. Taking all these findings together, it can be assumed that NO and IgE are closely related to cholinergic contraction in the trachea or to asthmatic reaction. Recently, Nakajima *et al.* (1993) have demonstrated that in mite-allergic asthma, Saiboku-to inhibits the induction of IgE-Fc epsilon receptor expression in the lymphocytes by mite allergen. Tamaoki *et al.* (1995a,b) have reported that Saiboku-to stimulates epithelial NO generation from cultured canine tracheal epithelium and from rabbit trachea in a concentration-dependent manner. These findings may indicate that Saiboku-to inhibits cholinergic bronchial contraction or asthmatic response by inhibiting expression of IgE receptor and NO generation.

Thus, Saiboku-to possesses various effects for protection from asthma, inflammatory disease, and neurotransmission induced by allergic reaction. These mechanisms, including inhibitory effects on histamine release from mast cells, the infiltration of eosinophils and T lymphocytes, IgE receptor expression, and a facilitatory effect on NO generation, might be interconnected. Further study is necessary to elucidate their relationship.

4.2.2.2 Anxiolytic effect

Histamine is widely distributed in the mammalian central nervous system (Hough, 1988; Panula *et al.*, 1989; Inagaki *et al.*, 1988, 1990; Yamatodani *et al.*, 1991). Brain histamine is localized in both histamine neurons and non-neuronal mast cells (Yamatodani *et al.*, 1982; Goldschmidt *et al.*, 1985; Onoue *et al.*, 1993; Bugajski *et al.*, 1995). The involvement of the histaminergic system in multifarious brain functions and various behaviors has been reviewed (Yamatodani *et al.*, 1991; Onodera and Miyazaki, 1999). With regard to the relationship between cerebral histamine and anxiety, clinically effective anxiolytic drugs—diazepam, a benzodiazepine (Oishi *et al.*, 1986; Chikai *et al.*, 1993), and buspirone, a serotonin ($5HT_{1A}$) agonist (Oishi *et al.*, 1992)—have been found to decrease the turnover rate of brain histamine in rodents. These findings suggest that cerebral histamine may play an important role in the regulation of anxiety.

Regarding the anxiolytic effect of Saiboku-to, Watanabe *et al.* (1973, 1983) have reported that Magnoliae Cortex, a constituent herb of Saiboku-to, possesses a central depressant effect. Recently, Kuribara *et al.* (1996) have demonstrated an anxiolytic-like effect of Saiboku-to using an elevated plus-maze test in mice. However, the anxiolytic mechanism of Saiboku-to remains unclear.

Imaizumi and Onodera (1993) and Imaizumi *et al.* (1996) have shown that anxiety-like behavioral activity is induced by the combination of thioperamide (a neuronal histamine releaser having the inhibitory effect of histamine H_3 autoreceptors (Arrang *et al.*, 1985; Mochizuki *et al.*, 1991)) with zolanitidine (a histamine H_2 receptor antagonist). In addition, Yuzurihara *et al.* (2000a) have demonstrated that anxiety-like behavioral activity is also induced by co-injection of Compound 48/80 (a non-neuronal selective mast cell histamine releaser (Russell *et al.*, 1990; Wu *et al.*, 1993; Ikarashi *et al.*, 2001b)) with cimetidine (a histamine H_2 receptor antagonist), using a light/dark test. As Saiboku-to has the potential to inhibit histamine release (see previous section), these neuronal and non-neuronal histaminergic-induced anxiety models in mice are

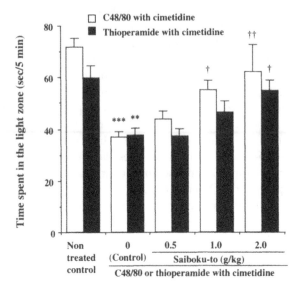

Figure 4.1 Effect of 7-day administration of Saiboku-to on histaminergic-induced anxiety-like behavioral activity (time spent in the light zone) in the light/dark test. Saiboku-to (0.5–2.0 mg/kg, p.o.) was administered once a day for 7 days in mice. The light/dark test was carried out 24 h after the last administration of Saiboku-to. Anxiety was induced by co-injection of either Compound 48/80 (C48/80: 1.0 µg/2 µl, i.c.v.) or thioperamide (10.0 mg/kg, i.p.) with cimetidine (10.0 µg/2 µl, i.c.v.) 60 min before the light/dark test. The detailed procedure is described in Yuzurihara *et al.* (2000a,b). Data are expressed as mean ± SE ($n = 10$). Statistical significance was evaluated by a one-way ANOVA, followed by Dunnet's least significant difference procedure: ** $p < 0.01$ and *** $p < 0.001$ compared to non-treated control group, and † $p < 0.05$ and †† $p < 0.01$ compared to C48/80 or thioperamide with cimetidine-treated control group.

useful for assessing the effect of Saiboku-to on the brain histaminergic system in a state of anxiety.

We have investigated the mechanism of action of Saiboku-to as an anxiolytic drug in mice (Yuzurihara *et al.*, 2000b) using two anxiogenic protocols: (i) targeting mast cells; injecting mice with Compound 48/80 in combination with cimetidine; and (ii) targeting neuronal cells; treating mice with thioperamide in combination with cimetidine. The results are shown in Figure 4.1. Both Compound 48/80 and thioperamide induced a marked and significant decrease in the time spent in the light zone on co-injection with cimetidine, in comparison with corresponding non-treated control group. These histaminergic-induced changes were significantly antagonized or inhibited by pretreatment with Saiboku-to for 7 days. The shortening on prolongation of time was not due to behavioral toxicity such as motor incoordination or activation of motor behavior, since locomotory activity was not different from that of the control mice. In a light/dark test, anxiolytics were found to prolong the time spent in the light zone, while anxiogenics decrease this parameter (De Angelis, 1995; Imaizumi *et al.*, 1994). Therefore, taken together, our results suggest not only that Saiboku-to inhibits histaminergic-induced experimental anxiety but also that the anxiolytic properties of Saiboku-to are closely related to the histaminergic system in the brain.

We assume that Saiboku-to may inhibit a common factor in histamine secretion induced by both Compound 48/80 and thioperamide. Ca^{2+}-dependent exocytosis (Starke *et al.*, 1989) is thought to be one of the candidates for their common mechanism. Saiboku-to might inhibit histamine release by interfering with the influx of Ca^{2+} into the mast cells and histamine nerve terminals, although further studies are necessary to confirm this hypothesis.

Furthermore, a 7-day pretreatment with Saiboku-to enhances the anxiolytic-like effect of diazepam in mice, as suggested by the results of an elevated plus-maze test (Kuribara *et al.*, 1996). Kuribara *et al.* (1996) have suggested that the anxiolytic-like effect of Saiboku-to might be related to $GABA_A$ receptors, as the effect is enhanced by $GABA_A$ agonist diazepam and antagonized by the antagonist flumazenil. However, Oishi *et al.* (1986) have demonstrated, using mouse brain, that diazepam produces an inhibitory effect on histamine turnover by acting on $GABA_A$ receptors. Chikai *et al.* (1993) have also reported that several sedative drugs, $GABA_A$ receptor agonists, such as muscimol, diazepam and pentobarbital, inhibit histamine release in rat striatum. Takeda *et al.* (1984) and Airaksinen *et al.* (1992) suggest the coexistence of glutamate decarboxylase and histidine decarboxylase in neurons in the hypothalamus. Thus, it is possible that GABA released from these neurons may regulate histaminergic activity of the same neurons; or it is likely that $GABA_A$ receptors in histaminergic nerve terminals are involved in the regulation of histamine release. Enhancement of the anxiolytic effect of diazepam by the administration of Saiboku-to might be due to synergis, with different mechanisms causing the inhibition of histamine release: direct inhibition of histamine release by Saiboku-to and the indirect inhibition by diazepam via $GABA_A$ receptors.

As a traditional oriental herbal medicine, Saiboku-to has for a long time been used empirically in the psychiatric field for treatment of anxiety-related disorders such as neurosis. Our present study in which experimentally induced anxieties were inhibited by subacute treatment with Saiboku-to in mice would support clinical application of Saiboku-to as anxiolytics. It seems likely that the anxiolytic-like effect of Saiboku-to is closely related to the inhibition of histamine release in the brain.

4.2.2.3 *Anti-ulcer effect*

Neurosis and psychosomatic diseases are thought to be caused mainly by a variety of physical stresses and mental anxieties (Adams and Victor, 1993; Kudo and Kudo, 1995). In particular, the stomach, duodenum and large intestine respond to stress more acutely, resulting in peptic ulcer (Varis, 1987; Hernandez *et al.*, 1993; Piper and Tennant, 1993). For this reason, anxiolytic drugs have been used clinically in the treatment of patients with acid-peptic diseases (File and Pearce, 1981; Imperato *et al.*, 1994; Sieghart, 1995) in a manner similar to the use of anti-ulcer drugs, including histamine H_2 receptor antagonists, anti-cholinergics and proton pump inhibitors (Isenberg *et al.*, 1991). It is well known that histamine is one of the neurotransmitters regulating gastric acid secretion, an aggressive factor affecting the onset of gastric ulcer. It was suggested in the previous sections that the anti-allergic and anxiolytic effects of Saiboku-to might be related to the inhibition of histamine release. Therefore, Saiboku-to may be expected to protect the induction of gastric ulcer if it does indeed possess anti-histamine release effect. Yuzurihara *et al.* (1999) have demonstrated that Saiboku-to inhibited the development of gastric ulcer induced by restraint water-immersion stress: the lesion index, calculated as the cumulative length of gastric lesion in the gross appearance of the stomach, is dose-dependently inhibited by oral administration of Saiboku-to 30 min before applying the stress. It has been reported that stress-induced gastric lesion develops as a result of vagal nerve stimulation, which increases gastric secretion (Kitagawa *et al.*, 1979) and gastric motility (Garrick *et al.*, 1986), causes the diminution of gastric mucus (Hakkinen *et al.*, 1966), and alters the microcirculation of the gastric mucosa (Guth and Hall, 1966). The anticholinergic agent atropine, the histamine H_2 inhibitor cimetidine, and the anxiolytic drug diazepam inhibit the stress-induced gastric acid, whereas cetraxate and sucralfate, as agents enhancing mucosal defensive factors, do not (Matsuda *et al.*, 1996; Yuzurihara *et al.*, 1999). These findings imply that restraint water-immersion stress-induced gastric lesions arise primarily as a result of gastric acid secretion, an aggressive factor affecting the induction of gastric ulcer. In pylorus-ligated rats, intraduodenal Saiboku-to inhibits the volume, the acid output and the acidity (the inhibited acidity is also supported by increased pH) of gastric juice (Yuzurihara *et al.*, 1999). In their study, as Saiboku-to was injected into the duodenum following pylorus ligation, the inhibition of the gastric secretion induced by Saiboku-to is thought to be due to absorption from the duodenum. However, many factors are thought to be responsible for acid inhibition. They include (i) anxiolytic effect in the central nervous system (CNS); (ii) anti-vagal nerve activity; (iii) inhibition of cholinergic, histaminergic and gastrinergic neurotransmissions; (iv) inhibition of the activities of various post-synaptic receptors such as histamine H_2, muscarine and gastrin receptors located on the basolateral membrane of the acid-secreting parietal cell; and (v) inhibition of the activity of final common acid-secreting proton pump, H^+/K^+-ATPase located on the canalicular surface of the cell (Isenberg *et al.*, 1991). Thus, to clarify the effect of the inhibitory mechanism of Saiboku-to on gastric acid secretion, we used vagotomized pylorus-ligated rats (Ikarashi *et al.*, 2001a). The use of vagotomized rats eliminated the effect of Saiboku-to on the efferent activity in gastric vagus via CNS. Under vagotomy, exogenous injection of cimetidine, histamine H_2 receptor antagonist, or 16,16-dimethyl prostaglandin E_2 (dmPGE$_2$) inhibited gastric acid secretion stimulated by the secretagogues bethanechol and tetragastrin as well as histamine. In addition, atropine, a muscarinic receptor inhibitor, and proglumide, a gastrinic receptor inhibitor, inhibited gastric acid secretion stimulated by the respective secretagogues bethanechol or tetragastrin, but did not antagonize the acid secretion stimulated by other secretagogues. It has been demonstrated that (i) histamine increases gastric acid secretion by binding to a stimulatory membrane receptor (R_s or H_2 receptor) that is coupled via a stimulatory GTP-binding protein (G_s), which in turn facilitates adenylate cyclase activity and the level of intracellular cAMP; and (ii) prostaglandin E decreases gastric acid secretion by binding to an inhibitory membrane receptor (R_i) that is coupled via an inhibitory GTP-binding protein (G_i), which in turn inhibits adenylate cyclase activity and the level of cAMP (Shamburek and Schubert, 1993). Taking all our results together, it can be suggested that not only are the receptors for histamine, ACh and gastrin present in the parietal cell, but also ACh and gastrin are capable of releasing histamine from mucosal histamine-storing cells such as enterochromaffin-like and mast cells (Hakanson and Sundler, 1987). Our results also suggest that histamine plays a pivotal role in stimulation of the parietal cell, and support the hypothesis advanced by Hirschowitz *et al.* (1995).

In the vagotomized pylorus-ligated rats, Saiboku-to inhibited gastric acid secretion induced by both bethanechol and tetragastrin, but not by histamine. This result is in

Table 4.3 Effect of Saiboku-to on gastric acid, histamine and acetylcholine (ACh) outputs into gastric juice in pylorus-ligated rats

Group	Gastric acid (μEq/h)	Histamine (ng/h)	ACh (pmol/h)
Control	255 ± 16	119 ± 10	102 ± 20
0.25 g/kg Saiboku-to	209 ± 16	101 ± 9	96 ± 30
0.50 g/kg Saiboku-to	106 ± 27***	76 ± 9*	104 ± 29
1.00 g/kg Saiboku-to	68 ± 9***	49 ± 9***	99 ± 21

Animals were killed 4 h after pylorus ligation for gastric juice collection. Saiboku-to (0.25, 0.50 and 1.00 g/kg) or vehicle (10 ml/kg water in control) was injected intradermally immediately after pylorus ligation. In the experiment for the determination of ACh level, 1 ml of 1 mM eserine was orally administered after pylorus ligation. For details of experimental procedures, see Ikarashi *et al.* (2000, 2001a). Data are expressed as mean ± SE ($n = 10$). Statistical significance was assayed by a one-way ANOVA, followed by Bonferroni multiple-comparison test: * $p < 0.05$ and *** $p < 0.001$ compared to control group.

line with those obtained by the investigation of the effect of certain mast cell stabilizers such as sodium chromoglycate and FPL52694 (Nicol *et al.*, 1981) on gastric acid secretion in rats. Tabuchi and Furuhama (1994) have also demonstrated that DS-4574, a mast cell stabilizer, inhibits not only gastric acid secretion induced by both carbachol and pentagastrin (but not by histamine), but also histamine leakage into the gastric juice produced by carbachol or pentagastrin. This suggests that the inhibitory effect of these drugs on gastric acid secretion is mediated by inhibition of endogenous histamine release from histamine-storing cells in the stomach. Recently, we demonstrated that histamine and ACh levels in gastric juice of pylorus-ligated rat reflected the activities of histaminergic and cholinergic neurotransmitters (Ikarashi *et al.*, 2000). Table 4.3 shows that Saiboku-to inhibits gastric acid and histamine outputs, without affecting ACh output into gastric juice, suggesting that Saiboku-to may inhibit histamine release from histamine-storing cells in the stomach. These results also support the hypothesis that inhibited histamine release is responsible for inhibition of gastric acid secretion by Saiboku-to.

Saiboku-to consists of ten medical herbs. Magnolol in Magnoliae Cortex inhibits stress-induced lesions and gastric secretion (Yano, 1997). It is also known that glycyrrhizin in Glycyrrizae Redix and its derivative, carbenoxolone, protect restraint water-immersion stress-induced ulcer (Yano, 1997). Polysaccharides prepared from Bupleuri Radix have been reported to protect gastric mucosa (Yamada *et al.*, 1991). Thus, some constituent ingredients in Saiboku-to are likely agents responsible for the anti-histamine effect.

4.2.2.4 *Anti-histamine release effect*

The anti-allergic, anxiolytic and anti-ulcer effects of Saiboku-to have been described in previous sections. The evidence presented suggests that various effects of Saiboku-to are closely related to the inhibition of histamine release from biogenic mast cells. However, the relative importance of the ten constituent herbs in the effect of Saiboku-to remains to be fully demonstrated. Thus, in order to elucidate the anti-histamine release effect of Saiboku-to, we investigated the effects of the constituent herbs on histamine release induced by a mast cell degranulator, Compound 48/80, from peritoneal mast cells in rats (Ikarashi *et al.*, 2001b). The effects of various concentrations of

Table 4.4 Inhibitory effect of Saiboku-to on histamine release induced by Compound 48/80 from peritoneal mast cells in rats

Group	% Release	Inhibition rate (%)
Compound 48/80 (C48/80)	33.5 ± 3.0	0 (Control)
0.25 mg/ml Saiboku-to + C48/80	20.7 ± 2.6**	38.2
0.50 mg/ml Saiboku-to + C48/80	13.8 ± 2.1***	58.8
1.00 mg/ml Saiboku-to + C48/80	8.3 ± 1.7***	75.2

50 µl of Saiboku-to (final concentration: 0.25, 0.50 and 1.00 mg/ml) was added to 400 µl suspension of rat peritoneal mast cells adjusted to 1×10^5 cells/ml. After incubation at 37 °C for 10 min, histamine release from the mast cells was elicited by addition of 50 µl of 0.6 µg/ml Compound 48/80 (C48/80) to Saiboku-to-pretreated cell suspension. For details see Ikarashi *et al.* (2001b). The total histamine level in mast cells used in the present study was 1138 ± 39 ng. Histamine release was expressed as a percentage of the total histamine level. Basal histamine release was 4.37% ± 0.37% of the total histamine level. The inhibition rate of each concentration of Saiboku-to was expressed as a percentage of the C48/80-induced histamine release level. Data are expressed as mean ± SE (*n* = 6). Statistical significance was assayed by a one-way ANOVA, followed by Bonferroni multiple-comparison test: ** $p < 0.01$ and *** $p < 0.001$ compared to the C48/80-treated control group.

Saiboku-to (0.25–1.0 mg/ml) on Compound 48/80-induced histamine release from mast cells are shown in Table 4.4. The percentage of histamine release from mast cells induced by 0.6 µg/ml of Compound 48/80 was 33.5 ± 3.0% of the total histamine level (1138 ± 39 ng). Saiboku-to inhibited Compound 48/80-induced histamine release in a concentration-dependent manner. Significant inhibition was observed at concentrations of Saiboku-to exceeding 0.25 mg/ml (−38.2%, $p < 0.01$). The highest dose used in this study, 1.0 mg/ml, inhibited 75.2% of the Compound 48/80-induced histamine release. These results clearly indicate that Saiboku-to inhibits Compound 48/80-induced histamine release from peritoneal mast cells in rats. These biochemical results are also supported by morphological findings on peritoneal mast cells in mice (Toda *et al.*, 1988) and rats (Ikarashi *et al.*, 2001b).

The inhibitory effect of Saiboku-to on the underlying histamine release mechanism suggests that at least some of the constituent herbs of Saiboku-to may play a role in the inhibitory effect. Together with Saiboku-to, Syo-saiko-to and Dai-saiko-to belong to the group of saiko-agents. Toda *et al.* (1987) have reported that Syo-saiko-to and Dai-saiko-to have inhibitory effects on degranulation of and histamine release from peritoneal mast cells in mice induced by Compound 48/80. As both saiko-agents contain Bupleuri Radix, Pinelliae Tuber, Scutellariae Radix, Zizyphi Fructus and Zingiberis Rhizome, the authors suggest that these constituent herbs may possess the inhibitory effect. However, in the screening test to examine the inhibitory effect of histamine release among the ten constituent herbs (Ikarashi *et al.*, 2001b), four herbs were found to possess a potent effect: Magnoliae Cortex, Perillae Herba, Bupleuri Radix and Hoelen, as shown in Table 4.5. No significant change was produced by the other six herbs. In dose–response studies on the four active herbs, dose-dependent inhibitory responses versus the logarithm of the drug concentration were found in four herbs (Ikarashi *et al.*, 2001b). The correlative formulas of the logarithmic linearity and the values of 50% inhibitory concentration (IC_{50}) for the four herbs are shown in Table 4.6. The potency was highest in Magnoliae Cortex (IC_{50} = 56.8 µg/ml), followed by Perillae Herba (IC_{50} = 175.8 µg/ml), Bupleuri Radix (IC_{50} = 356.5 µg/ml), and Hoelen (IC_{50} = 594.3 µg/ml). As shown in Table 4.4, 1.0 mg/ml of Saiboku-to

Table 4.5 Effect of the ten constituent herbs in Saiboku-to on Compound 48/80-induced histamine release from peritoneal mast cells in rats

Constituent herb	Inhibition rate (%)	Constituent herb	Inhibition rate (%)
Magnoliae Cortex	93.00 ± 0.55***	Glycyrrhizae Radix	9.10 ± 4.72
Perillae Herba	57.84 ± 8.06**	Zizyphi Fructus	8.91 ± 5.08
Bupleuri Radix	30.63 ± 5.18*	Penelliae Tuber	6.35 ± 6.49
Hoelen	25.73 ± 5.11*	Scutellariae Radix	−3.28 ± 5.27
Ginseng Radix	9.34 ± 4.64	Zingiberis Rhizoma	−3.81 ± 3.60

The effect of all herbs on the histamine release was evaluated at a fixed concentration of 0.25 mg/ml. The inhibition rate of each herb was expressed as a percentage of the Compound 48/80 (0.6 μg/ml)-induced histamine release level. The Compound 48/80-induced histamine release level was 30.0% ± 3.3% of the total histamine level (1082.5 ± 53.2 ng) in this study. The outline of the experimental procedure can be found in Table 4.4. Data are expressed as mean ± SE ($n = 6$). Statistical significance was assayed by a one-way ANOVA, followed by Bonferrroni multiple-comparison test: * $p < 0.05$, ** $p < 0.01$ and *** $p < 0.001$ compared to Compound 48/80-induced histamine release level.

Table 4.6 Dose–response correlative formulas and 50% inhibitory concentration (IC_{50}) of Magnoliae Cortex, Perillae Herba, Bupleuri Radix and Hoelen for Compound 48/80-induced histamine release level from peritoneal mast cells in rats

Herbs	Correlative formula	r	IC_{50} (μg/ml)
Magnoliae Cortex	$y = 131.7 + 65.6 \log(x)$	0.978	56.8
Perillae Herba	$y = 95.6 + 60.4 \log(x)$	0.992	175.8
Bupleuri Radix	$y = 77.7 + 61.8 \log(x)$	0.983	356.5
Hoelen	$y = 64.5 + 64.1 \log(x)$	0.998	594.3

y = inhibition rate (%); x = drug concentration (mg/ml); r = correlation coefficient.

inhibited 75% of Compound 48/80-induced histamine release. This concentration of Saiboku-to (1.0 mg/ml) contains 88.2 μg of Magnoliae Cortex, 58.8 μg of Perillae Herba, 205.9 μg of Bupleuri Radix and 147.1 μg of Hoelen; the inhibition rates of histamine release for these concentrations from the formulas in Table 4.6 were estimated to be 62.68% for Magnoliae Cortex, 21.00% for Perillae Herba, 35.24% for Bupleuri Radix, and 11.15% for Hoelen. These results suggest that the anti-histamine releasing effect of Saiboku-to is mainly due to the effect of Magnoliae Cortex and the synergism of the three other herbs. As described above, some of the other herbs contained in Saiboku-to may possess anti-histamine release effect; however, their potency is likely to be less than that of the four herbs, as shown in Table 4.5.

Compound 48/80, a condensation product of N-methyl-p-methoxyphenethylamine with formaldehyde, comprises a family of cationic amphiphiles of various degrees of polymerization (Gietzen et al., 1983). This substance has been widely used as a selective histamine release agent or degranulator from mast cells in rats and mice (Toda et al., 1988; Wu et al., 1993). Recently, it has been reported that the mechanism of histamine release from mast cells induced by Compound 48/80 is closely related to activation of phosphatidylinositol (PI) turnover and Ca^{2+} mobilization through guanine nucleotide-binding regulatory protein (Wu et al., 1993; Kamei et al., 1997; Barrocas et al., 1999). Saiboku-to suppresses antigen-induced influx of Ca^{2+} into the mast cells of mice (Dobashi et al., 1993). The biphenyl ingredients magnolol and honokiol in the

Magnoliae Cortex in Saiboku-to antagonize voltage-sensitive Ca^{2+} channels (Yamahara *et al.*, 1986; Teng *et al.*, 1990). Thus, Saiboku-to might inhibit histamine release by interfering with the influx of Ca^{2+} into the mast cells.

4.2.3   Conclusion

We have described recent findings related to Saiboku-to including our results on allergy, anxiety and ulcer. Scientific data suggest that various effects of Saiboku-to seem to be related to the inhibition of biogenic responses induced by histamine. We suggest that the herb chiefly responsible for the anti-histamine releasing effect of Saiboku-to may be Magnoliae Cortex. The exact nature of the anti-histamine release effect of the active ingredients of Magnoliae Cortex remains to be investigated. However, various ingredients have been isolated from Magnoliae Cortex and identified: β-eudesmol, α- and β-pinenes, camphene and limonene as the essential oils; magnocurarine and magnoflorine as the alkaloids; and magnolol and honokiol as the biphenyl compounds (Pu *et al.*, 1990; Sashida, 1994; Maruyama *et al.*, 1998). In particular, it has been demonstrated that magnolol and honokiol exert the central depressant effect (Watanabe *et al.*, 1983). Recently, Maruyama *et al.* (1998) reported that both ingredients are anxiolytic agents. As described earlier, experimental anxiety is induced by the administration of histaminergics, and the anxiety is inhibited by pretreatment with Saiboku-to (Yuzurihara *et al.*, 2000a,b). Saiboku-to suppresses antigen-induced influx of Ca^{2+} into the mast cells in mice (Dobashi *et al.*, 1993). Furthermore, it has been suggested that the biphenyl ingredients antagonize voltage-sensitive Ca^{2+} channels (Yamahara *et al.*, 1986; Teng *et al.*, 1990). Thus, biphenyl compounds magnolol and honokiol are likely to be the active compounds responsible for the anti-histamine release effect of Saiboku-to, although the precise effects of these ingredients on histamine release from mast cells require further investigation.

We hope this review lay the foundation for further investigation to assess the mechanism underlying the effects of Saiboku-to.

References for section II

Adams, R.D. and Victor, M. (1993) Psychiatric disorders. In *Principles of Neurology*, 5th edn, edited by R.D. Adams and M. Victor, pp. 1285–1310. New York: McGraw-Hill.

Airaksinen, M.S., Alanen, S., Szabat, E., Visser, T.J. and Panula, P. (1992) Multiple neurotransmitters in the tuberomammillary nucleus: comparison of rat, mouse, and guinea-pig. *J. Comp. Neurol.*, 323, 103–116.

Arrang, J.M., Garbarg, M. and Schwartz, J.C. (1985) Autoregulation of histamine release in brain by presynaptic H_3 receptors. *Neuroscience*, 15, 553–562.

Barrocas, A.M., Cochrane, D.E., Carraway, R.E. and Feldberg, R.S. (1999) Neurotensin stimulation of mast cell secretion is receptor-mediated, pertussis-toxin sensitive and requires activation of phospholipase C. *Immunopharmacology*, 41, 131–137.

Bugajski, A.J., Chlap, Z., Bugajski, J. and Borycz, J. (1995) Effect of Compound 48/80 on mast cells in brain structures and on corticosterone secretion. *J. Physiol. Pharmacol.*, 46, 513–522.

Chikai, T., Oishi, R. and Saeki, K. (1993) Microdialysis study of sedative drugs on extracellular histamine in the striatum of freely moving rats. *J. Pharmacol. Exp. Ther.*, 266, 1277–1281.

De Angelis, L. (1995) Effects of valproate and lorazepam on experimental anxiety: tolerance, withdrawal, and role of clonidine. *Pharmacol. Biochem. Behav.*, 52, 329–333.

Dobashi, K., Watanabe, K., Kobayashi, S., Mori, M. and Nakazawa, T. (1993) Suppression of Saiboku-to of antigen-induced Ca^{2+} influx into mast cells. *Kampo and Immuno-allergy*, 7, 29–36.

File, S.F. and Pearce, J.B. (1981) Benzodiazepines reduced gastric ulcers induced in rat by stress. *Br. J. Pharmacol.*, 74, 593–599.

Garrick, T., Leung, F.W., Buack, S., Hirabayashi, K. and Guth, P.H. (1986) Gastric motility is stimulated but overall blood flow is unaffected during cold restraint in the rat. *Gastroenterology*, 91, 141–149.

Gibson, A., Mirzazadeh, S., Hobbs, A.J. and Moore, P.K. (1990) L-N^G-monomethyl arginine and L-N^G-nitro arginine inhibit non-adrenergic, non-cholinergic relaxation of the mouse anococcygeus muscle. *Br. J. Pharmacol.*, 99, 602–606.

Gietzen, K., Adamczyk-Engelmann, P., Wuthrich, A., Konstantinova, A. and Bader, H. (1983) Compound 48/80 is a selective and powerful inhibitor of calmodulin-regulated functions. *Biochim. Biophys. Acta.*, 736, 109–118.

Goldschmidt, R.C., Hough, L.B. and Glick, S.D. (1985) Rat brain mast cells: contribution to brain histamine levels. *J. Neurochem.*, 44, 1943–1947.

Guth, P.H. and Hall, P. (1966) Microcirculatory and mast cell changes in restraint induced gastric ulcer. *Gastroenterology*, 50, 562–570.

Hakanson, R. and Sundler, F. (1987) Localization of gastric histamine immunocytochemical observations. *Med. Biol.*, 65, 1–7.

Hakkinen, I., Hartiala, K. and Lang, H. (1966) The effect of restraint on the content of acid polysaccharides of glandular gastric wall in rat. *Acta Physiol. Scand.*, 66, 333–336.

Hernandez, D.E., Arandia, D. and Dehesa, M. (1993) Role of psychosomatic factors in peptic ulcer disease. *J. Physiol. Paris*, 87, 223–227.

Hirschowitz, B.I., Keeling, D., Lewin, M., *et al.* (1995) Pharmacological aspects of acid secretion. *Dig. Dis. Sci.*, 40, 3–23.

Homma, M., Oka, K., Kobayashi, H., *et al.* (1993) Impact of free magnolol excretions in asthmatic patient who responded well to Saiboku-to a Chinese herbal medicine. *J. Pharm. Pharmacol.*, 45, 844–846.

Homma, M., Oka, K., Niitsuma, T. and Itoh, H. (1994) A novel 11β-hydroxysteroid dehydrogenase inhibitor contained in Saiboku-to, a herbal remedy for steroid-dependent bronchial asthma. *J. Pharm. Pharmacol.*, 46, 305–309.

Hough, L.B. (1988) Cellular localization and possible functions for brain histamine: recent progress. *Prog. Neurobiol.*, 30, 469–505.

Ichinose, M., Miura, M., Tomaki, M., *et al.* (1996) Incubation with IgE increases cholinergic neurotransmission in human airway *in vitro*. *Am. J. Respir. Crit. Care. Med.*, 154, 1272–1276.

Ikarashi, Y., Takahashi, A., Ishimaru, H., Shiobara, T. and Maruyama, Y. (1998) The role of nitric oxide in striatal acetylcholine release induced by N-methyl-D-aspartate. *Neurochem. Int.*, 33, 255–261.

Ikarashi, Y., Yuzurihara, M., Shinoda, M. and Maruyama, Y. (2000) Effect of 2-deoxy-D-glucose on acetylcholine and histamine levels in gastric juice of pylorus-ligated rats anesthetized with urethane. *J. Chromatogr. B*, 742, 295–301.

Ikarashi, Y., Yuzurihara, M. and Maruyama, Y. (2001a) Inhibition of gastric acid secretion by Saiboku-to, an oriental herbal medicine, in rats. *Dig. Dis. Sci.*, 46, 997–1003.

Ikarashi, Y., Yuzurihara, M., Tokieda, T., *et al.* (2001b) Effects of an oriental herbal medicine "Saiboku-to" and its constituent herbs on Compound 48/80-induced histamine release from peritoneal mast cells in rats. *Phytomedicine*, 8, 8–15.

Imaizumi, M. and Onodera, K. (1993) The behavioral and biochemical effects of thioperamide, a histamine H_3-receptor antagonist, in a light/dark test measuring anxiety in mice. *Life Sci.*, 53, 1675–1683.

Imaizumi, M., Suzuki, T., Machida, H. and Onodera, K. (1994) A full automated apparatus for a light/dark test measuring anxiolytic or anxiogenetic effects of drugs in mice. *Jpn. J. Psychopharmacol.*, 14, 83–91.

Imaizumi, M., Miyazaki, S. and Onodera, K. (1996) Effects of betahistine, a histamine H_1 agonist and H_3 antagonist, in a light/dark test in mice. *Methods Find. Exp. Clin. Pharmacol.*, 18, 19–24.

Imperato, A., Dazzi, L., Obinu, M.C., Gessa, G.L. and Biggio, G. (1994) The benzodiazepine receptor antagonist flumazenil increases acetylcholine release in rat hippocampus. *Brain Res.*, 647, 167–171.

Inagaki, N., Yamatodani, A., Ando-Yamamoto, M., Tohyama, M., Watanabe, T. and Wada, H. (1988) Organization of histaminergic fibers in the rat brain. *J. Comp. Neurol.*, 273, 283–300.

Inagaki, N., Toda, K., Taniuchi, I., *et al.* (1990) An analysis of histaminergic efferents of the tuberomammillary nucleus to the medial preoptic area and inferior colliculus of the rat. *Exp. Brain Res.*, 80, 374–380.

Isenberg, J.I., McQuaid, K.R., Laine, L. and Rubin, W. (1991) Acid-peptic disorders. In *Textbook of Gastroenterology*, edited by T. Yamada, D.H. Alpers, C. Owyang, D.W. Powell and F.E. Silverstein, pp. 1241–1339. Philadelphia: J.B. Lippincott.

Kamei, C., Mio, M., Yoshida, T., Saito, Y., Toyoda, Y. and Tsuriya, Y. (1997) Effect of active metabolite of the antiallergic agent tazanolast on histamine release from rat mast cells. *Arzneim.-Forsch./Drug Res.*, 47, 183–186.

Kitagawa, H., Fujiwara, M. and Osumi, Y. (1979) Effects of water-immersion stress on gastric secretion and mucosal blood flow in rats. *Gastroenterology*, 77, 298–302.

Koda, A., Nishiyori, T., Nagai, H., Matsuura, N. and Tsuchiya, H. (1982) Anti-allergic actions of crude drugs and blended Chinese traditional medicines: effects on Type I and Type IV allergic reactions. *Folia Pharmacol. Jpn.*, 80, 31–41.

Kudo, Y. and Kudo, T. (1995) Recent progress in development of psychotropic drug:anti-anxiety drugs. *Jpn. J. Neuropsychopharmacol.*, 15, 75–86.

Kuribara, H., Morita, M., Ishige, A., Hayashi, K. and Maruyama, Y. (1996) Investigation of the anxiolytic effect of the extracts derived from Saiboku-to, an oriental herbal medicine, by an improved plus-maze test in mice. *Jpn. J. Neuropsychopharmacol.*, 18, 179–190.

Li, C.G. and Rand, M.J. (1989) Evidence for a role of nitric oxide in the neurotransmitter system mediating relaxation of the rat anococcygeus muscle. *Clin. Exp. Pharmacol. Physiol.*, 16, 933–938.

Li, C.G. and Rand, M.J. (1991) Evidence that part of the NANC relaxant response of guinea-pig trachea to electric field stimulation is mediated by nitric oxide. *Br. J. Pharmacol.*, 104, 565–569.

Maruyama, Y., Kuribara, H., Morita, M., Yuzurihara, M. and Weintraub, S.T. (1998) Identification of magnolol and honokiol as anxiolytic agents in extracts of Saiboku-to, an oriental herbal medicine. *J. Nat. Prod.*, 61, 135–138.

Matsuda, M., Kanita, R., Tsutsui, F. and Yamashita, A. (1996) Antiulcer properties of Sho-saiko-to. *Folia Pharmacol. Jpn.*, 108, 217–225.

Mochizuki, T., Yamatodani, A., Okakura, K., Takemura, M., Inagaki, N. and Wada, H. (1991) *In vivo* release of neuronal histamine in the hypothalamus of rats measured by microdialysis. *Naunyn-Schmiedebergs Arch. Pharmacol.*, 343, 190–195.

Nakajima, S., Tohda, Y., Ohkawa, K., Chihara, J. and Nagasawa, Y. (1993) Effect of Saiboku-to (TJ-96) on bronchial asthma: induction of glucocorticoid receptor, beta-adrenaline receptor, IgE-Fc epsilon receptor expression and its effect on experimental immediate and late asthmatic reaction. *Ann. NY Acad. Sci.*, 685, 549–560.

Nicol, A.K., Thomas, M. and Wilson, J. (1981) Inhibition of gastric acid secretion by sodium chromoglycate and FPL52694. *J. Pharm. Pharmacol.*, 33, 554–556.

Nishiyori, T., Tsuchiya, H., Inagaki, N., Nagai, H. and Koda, A. (1985) Effect of Saiboku-to, a blended Chinese traditional medicine, on Type I hypersensitivity reactions, particularly on experimentally-caused asthma. *Folia Pharmacol. Jpn.*, 85, 7–16.

Oishi, R., Nishibori, M., Itoh, Y. and Saeki, K. (1986) Diazepam-induced decrease in histamine turnover in mouse brain. *Eur. J. Pharmacol.*, 124, 337–342.

Oishi, R., Itoh, Y. and Saeki, K. (1992) Inhibition of turnover by 8-OH-DPAT, buspirone and 5-hydroxytryptophan in the mouse and rat brain. *Naunyn-Schmiedebergs Arch. Pharmacol.*, 345, 495–499.

Onodera, K. and Miyazaki, S. (1999) The role of histamine H_3 receptors in the behavioral disorders and neuropsychopharmacological aspects of its ligands in the brain. *Folia Pharmacol. Jpn.*, 114, 89–106.

Onoue, H., Maeyama, K., Nomura, S., *et al.* (1993) Absence of immature mast cells in the skin of Ws/Ws rats with a small deletion at tyrosine kinase domain of the c-kit gene. *Am. J. Pathol.*, 142, 1001–1007.

Panula, P., Pirvola, U., Auvinen, S. and Airaksinen, M.S. (1989) Histamine-immunoreactive nerve fibers in the rat brain. *Neuroscience*, 28, 585–610.

Piper, D.W. and Tennant, C. (1993) Stress and personality in patients with chronic peptic ulcer. *J. Clin. Gastroenterol.*, 16, 211–214.

Pu, Q.L., Pannell, L.K. and Xiao-duo, J. (1990) The essential oil of *Magnolia officinalis*. *Planta Med.*, 56, 129–130.

Russell, W.L., Henry, D.P., Phebus, L.A. and Clemens, J.A. (1990) Release of histamine in rat hypothalamus and corpus striatum *in vivo*. *Brain Res.*, 512, 95–101.

Sashida, Y. (1994) Chemical constituents of Magnoliae Cortex. *J. Tradit. Sino-Japanese Med.*, 15, 90–98.

Sekizawa, K., Fukushima, T., Ikarashi, Y., Maruyama, Y. and Sasaki, H. (1993) The role of nitric oxide in cholinergic neurotransmission in rat trachea. *Br. J. Pharmacol.*, 110, 816–820.

Shamburek, R.D. and Schubert, M.L. (1993) Pharmacology of gastric acid inhibition. *Baillière's Clin. Gastroenterol.*, 7, 23–54.

Sieghart, W. (1995) Structure and pharmacology of gamma-aminobutyric acid$_A$ receptor subtype. *Pharmacol. Rev.*, 47, 181–234.

Starke, K., Gothert, M. and Kilbinger, H. (1989) Modulation of neurotransmitter release by presynaptic autoreceptors. *Physiol. Rev.*, 69, 864–989.

Tabuchi, Y. and Furuhama, K. (1994) Inhibitory effect of DS-4574, a mast cell stabilizer with peptidoleukotriene receptor antagonism, on gastric acid secretion in rats. *Eur. J. Pharmacol.*, 255, 229–234.

Takeda, N., Inagaki, S., Shiosaka, S., *et al.* (1984) Immunohistochemical evidence for the coexistence of histidine decarboxylase-like and glutamate decarboxylase-like immunoreactivities in nerve cells of the magnocellular nucleus of the posterior hypothalamus of rats. *Proc. Natl. Acad. Sci. USA*, 81, 7647–7650.

Tamaoki, J., Takemura, H., Kondo, M. Chiyotani, A., Sakai, A. and Konno, K. (1995a) Effect of TJ-96, an anti-allergic herbal medicine, on tracheal transepithelial potential difference *in vivo. Res. Commun. Mol. Pathol. Pharmacol.*, 88, 39–50.

Tamaoki, J., Takemura, H., Kondo, M. Chiyotani, A. and Konno, K. (1995b) Effect of Saiboku-to, an antiasthmatic herbal medicine, on nitric oxide generation from cultured canine airway epithelial cells. *Jpn. J. Pharmacol.*, 69, 29–35.

Teng, C.M., Yu, S.M., Chen, C.C., Huang, Y.L. and Huang, T.F. (1990) EDRF-release and Ca^{2+}-channel blockade by magnolol, an antiplatelet agent isolated from Chinese herb Magnolia officinalis, in rat thoracic aorta. *Life Sci.*, 47, 1153–1161.

Toda, S., Kimura, M., Ohnishi, M. and Nakashima, K. (1987) Effects of Syo-saiko-to (Xiao-Chai-Hu-Tang), Dai-saiko-to (Da-Chai-Hu-Tang) and Saiko-ka-ryukotu-borei-to (Chai-Hu-Jia-Long-Gu-Mu-Li-Tang) on degranulation of and release of histamine from mouse peritoneal mast cells induced by Compound 48/80. *J. Med. Pharm. Soc. WAKAN-YAKU*, 4, 77–81.

Toda, S., Kimura, M., Ohnishi, M. and Nakashima, K. (1988) Effects of the Chinese herbal medicine "Saiboku-to" on histamine release from and the degranulation of mouse peritoneal mast cells induced by Compound 48/80. *J. Ethnopharmacol.*, 24, 303–309.

Tohda, Y., Haraguchi, R., Kubo, H., Muraki, M., Fukuoka, M. and Nakajima, S. (1999) Effects of Saiboku-to on dual-phase bronchoconstriction in asthmatic guinea-pigs. *Methods Find. Exp. Clin. Pharmacol.*, 21, 449–452.

Tucker, J.F., Brave, S.R., Charalambous, L., Hobbs, A.J. and Gibson, A. (1990) L-N^G-nitro arginine inhibits non-adrenergic, non-cholinergic relaxations of guinea-pig isolated tracheal smooth muscle. *Br. J. Pharmacol.*, 100, 663–664.

Umesato, Y. (1984) Asthmatic children and Chinese medicine. *Allergy*, 33, 1047–1053.

Varis, K. (1987) Psychosomatic factors in gastrointestinal disorders. *Ann. Clin. Res.*, 19, 135–142.

Watanabe, K., Goto, Y. and Yoshitomi, K. (1973) Central depressant effects of the extracts of Magnolia Cortex. *Chem. Pharm. Bull.*, 21, 1700–1708.

Watanabe, K., Watanabe, H., Goto, Y., Yamaguchi, M., Yamamoto, N. and Hagino, K. (1983) Pharmacological properties of magnolol and honokiol extracted from *Magnolia officinalis*: central depressant effects. *Planta Med.*, 49, 103–108.

Wu, C.Y., Chen, C.F. and Chiang, C.F. (1993) Stimulation of inositol phosphate production and GTPase activity by Compound 48/80 in rat peritoneal mast cells. *Biochem. Biophys. Res. Commun.*, 192, 204–213.

Yamada, H., Sun, X.B., Matsumoto, T., Ra, K.S., Hirano, M. and Kiyohara, H. (1991) Purification of anti-ulcer polysaccharides from the roots of *Bupleurum falcatum*. *Planta Med.*, 57, 555–559.

Yamatodani, A., Maeyama, K., Watanabe, T., Wada, H. and Kitamura, Y. (1982) Tissue distribution of histamine in a mutant mouse deficient in mast cells: clear evidence for the presence of non-mast-cell histamine. *Biochem. Pharmacol.*, 31, 305–309.

Yamatodani, A., Inagaki, N., Panula, P., Itowi, N., Watanabe, T. and Wada, H. (1991) Structure and functions of histaminergic neuron system. In *Handbook of Experimental Pharmacology*, vol. 97, edited by B. Uvnase, pp. 243–283. New York, Tokyo: Springer Verlag.

Yamahara, J., Miki, S., Matsuda, H. and Fujimura, H. (1986) Screening test for calcium antagonists in natural products: the active principle of *Magnolia obovata*. *Yakugaku Zasshi*, 106, 888–893.

Yano, S. (1997) Effect of kampo medicine on stress-induced ulceration. *Prog. Med.*, 17, 887–893.

Yuzurihara, M., Ikarashi, Y., Kase, Y., Torimaru, Y., Ishige, A. and Maruyama, Y. (1999) Effect of Saiboku-to, an oriental herbal medicine, on gastric lesion induced by restraint water-immersion stress or by ethanol treatment. *J. Pharm. Pharmacol.*, 51, 483–490.

Yuzurihara, M., Ikarashi, Y., Ishige, A., Sasaki, H., Kuribara, H. and Maruyama, Y. (2000a) Effects of drugs acting as histamine releasers or histamine receptor blockers on an experimental anxiety model in mice. *Pharm. Biochem. Behav.*, 67, 145–495.

Yuzurihara, M., Ikarashi, Y., Ishige, A., Sasaki, H. and Maruyama, Y. (2000b) Anxiolytic-like effect of Saiboku-to, an oriental herbal medicine, on histaminergics-induced anxiety in mice. *Pharmacol. Biochem. Behav.*, 67, 489–495.

SECTION III BEHAVIORAL DETERMINATION OF ANXIOLYTIC EFFECT OF HERBAL MEDICINES AS BEING CAUSED BY THE PRESCRIBED *MAGNOLIA* COMPONENT

Hisashi Kuribara

4.3.1 Introduction

Magnolia cortex is an important component of a number of crude drug prescriptions in oriental traditional medicines (Kampo and/or Wakan medicines). Some of the famous Kampo (or Wakan) medicines containing *Magnolia* cortex are presented in Table 4.7

Table 4.7 Well-known Kampo medicines prescribed with *Magnolia*

Name	Number of prescribed plants (Magnolia/total plants)	Clinical efficacy
Bukuryo-ingo-hange-koboku-to	10 (3.0/36.5 g)	Anxiety neurosis, gastritis, bronchitis, asthma
Bunsho-to	12 (2.0/23.5 g)	Dropsy, decreased urination
Choko-shitei-to	14 (1.0/22.4 g)	Hiccough
Dai-joki-to	4 (5.0/13.0 g)	Fever, acute pneumonia and hepatitis, hypertension, convulsion
Fukankin-shoki-san	8 (3.0/21.0 g)	Acute and chronic gastritis, decreased appetite, indigestion
Goshaku-san	17 (1.0/23.9 g)	Pain, gastrointestinal pain, headache, vegetative, dystonia
Hange-koboku-to	5 (3.0/20.0 g)	Anxiety neurosis, hysteria, insomnia
Heii-san	6 (3.0/14.0 g)	Decreased appetite, diarrhea, acute and chronic gastritis
Hochu-jishitsu-to	10 (2.0/24.5 g)	Dropsy, cirrhosis
Hoki-kenchu-to	9 (2.0/22.0 g)	Dropsy, cirrhosis
Irei-to	12 (2.5/25.0 g)	Diarrhea, abdominal pain, vomiting, acute gastrointestinal symptoms
Kakko-shoki-san	13 (2.0/22.5 g)	Headache, vomiting, diarrhea, acute gastrointestinal symptoms, abdominal pain
Kami-heii-san	8 (3.0/18.0 g)	Indigestion, stomach ache, decreased appetite
Koboku-mao-to	9 (4.0/41.0 g)	Cough, bronchitis, asthma
Koboku-sanmotsu-to	3 (5.0/10.0 g)	Abdominal pain, constipation
Kosha-heii-san	9 (3.0/20.5 g)	Indigestion, stomach ache, decreased appetite
Kosha-yoi-to	13 (2.0/24.7 g)	Stomach ache, chronic gastrointestinal symptoms
Koboku-shokyo-hange-ninjin-kanzo-to	5 (3.0/14.0 g)	Gastrointestinal catarrh
Mashinin-gan	6 (0.06/2.0 g)	Conspiration, hyperurination, decreased vitality
Saiboku-to	10 (3.0/37.0 g)	Asthma, cough, anxiety neurosis
Shishi-koboku-to	3 (4.0/9.0 g)	Insomnia
Shinpi-to	7 (3.9/20.0 g)	Asthma, bronchitis
Shobai-to	12 (2.9/24.9 g)	Vermifuge
Sho-joki-san	3 (3.0/7.0 g)	Constipation, hypertension
Soshi-koki-to	10 (2.5/22.0 g)	Bronchitis
Toki-to	10 (3.0/27.5 g)	Chill, abdominal pain
Toki-yoketsu-to	14 (1.5/26.0 g)	Vomiting, swallowing disturbance
Tsudo-san	10 (2.0/25.0 g)	Vegetative dystonia, constipation, bruising
Zyun-cho-to	11 (2.0/24.5 g)	Constipation

Modified from Nishimoto (1986), Namba and Tsuda (1998).

(Nishimoto, 1986; Namba and Tsuda, 1998). They have been used empirically for amelioration of a wide variety of clinical conditions such as abdominal pain, gastrointestinal disturbance, disturbed circulation, decreased appetite, asthma, thrombotic stroke, typhoid fever, fever, insomnia and anxiety (Juangsu New Medical College, 1985). Furthermore, some preclinical evaluations of Kampo medicines revealed that *Magnolia* extracts or its active chemicals, and *Magnolia*-prescribed Kampo medicines show various pharmacological actions such as sedative effect, central and/or peripheral muscle-relaxing effect, anticonvulsive effect, anti-ulcer effect, anti-vomiting effect, anti-allergic effect, anti-platelet aggregation effect, calcium channel blocking effect, suppression of the release of endothelium-derived relaxing factor, and antibacterial effect (Nishioka, 1983; Watanabe, 1994). In this section, the processes of determining the anxiolytic effect of *Magnolia*-prescribed Kampo medicines and identification of the active chemical(s) are reviewed.

4.3.2   Anxiolytic effect of Hange-koboku-to and Saiboku-to

Among the *Magnolia*-prescribed Kampo medicines, some such as Hange-koboku-to, Saiboku-to, Bukuryo-ingo-hange-koboku-to and Shishi-koboku-to have been considered to relieve anxiety, nervous tension and/or insomnia (Hosoya and Yamamura, 1988; Narita, 1990; Ishida *et al.*, 1999). Kuribara and Maruyama (1995, 1996) applied an elevated plus-maze test in mice to assess behaviorally whether Hange-koboku-to and Saiboku-to exhibited an anxiolytic effect. The effects of Yokukan-san (made of six plants) and Kami-kihi-to (made of 13 plants) were also assessed because both of these Kampo medicines have been used for amelioration of anxiety, neurosis and insomnia, although they do not contain *Magnolia*. Kakkon-to (made of seven plants) was administered as the Kampo medicine without anxiolytic effect.

4.3.2.1   *Plus-maze test*

The elevated plus-maze for mice used for the assessment of the anxiolytic effect of Kampo medicines was a slight modification of the apparatus initially designed for rats (Pellow *et al.*, 1985) and mice (Lister, 1987). The plus-maze consisted of four arms (6 cm × 30 cm, each) that extended from a central platform (8 cm × 8 cm) of nontransparent gray. Two of the four arms, namely the closed arms, had side-walls, and both the floors and side-walls were nontransparent gray. The other two arms, namely the open arms, had no side-walls and the floor was transparent, whereas the open arms of the originally designed apparatus were nontransparent. The plus-maze was set 40 cm above the base. For testing, a mouse was placed on the central platform, oriented randomly toward one of the closed arms. During the ensuring 5-min period, the cumulative time spent in the open arms was recorded. The mouse was considered to have entered either of the open arms when all four paws crossed the border between the central platform and the open arm.

4.3.2.2   *Activity test*

To estimate nonspecific effects of general activity on the plus-maze performance (Dawson and Tricklebank, 1995), the ambulatory activity of the mouse was measured for 5 min immediately after the plus-maze test with a tilting-type ambulometer equipped with bucket-like activity cage 20 cm in diameter (SMA-1: O'hara & Co., Tokyo). The apparatus detected slight tilts of the activity cage generated by the mouse's ambulation, i.e., horizontal movements of comparatively longer distance.

4.3.2.3 Anxiolytic effect

Since Kampo medicines have generally been administered for a certain period to develop the clinical efficacies, male mice of ddY strain were orally administered either *Magnolia*-prescribed Kampo medicines Hange-koboku-to (0.5–2 g/kg) or Saiboku-to (0.5–2 g/kg), or the *Magnolia*-nonprescribed Kampo medicines Yokukan-san (0.5–2 g/kg) or Kami-kihi-to (0.25–2 g/kg) once a day for 7 days, and the plus-maze test was conducted 24 h after the last administration. The results are shown in Figure 4.2. The mice treated with the four Kampo medicines exhibited significant prolongation in the time spent in the open arms without significant change in the ambulatory activity, indicating that the change in the plus-maze performance was caused by the anxiolytic effect. In contrast, the same treatment with Kakkon-to (0.5–2 g/kg), a Kampo medicine without any anxiolytic effect, did not affect the plus-maze performance or ambulatory activity. The benzodiazepine anxiolytic diazepam (0.5–1 mg/kg p.o.), singly administered 10 min before the plus-maze test, also dose-dependently

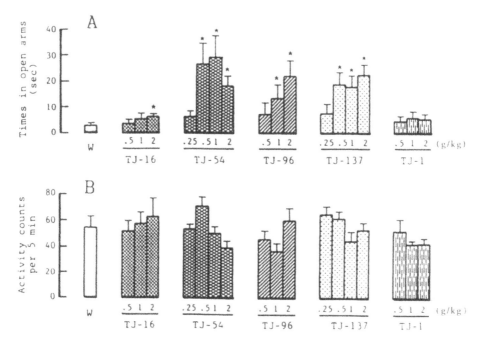

Figure 4.2 The effects of Hange-koboku-to (TJ-16), Yokukan-san (TJ-54), Saiboku-to (TJ-96), Kami-kihi-to (TJ-137) and Kakkon-to (TJ-1) assessed by elevated plus-maze (A) and activity tests (B) in mice. Mice were orally treated with either drug or water (control) once a day for 7 days, and the behavioral tests were carried out 24 h after the last drug administration. Data presented are mean with SEM of 10–12 mice. * $p < 0.05$, significant different as compared to the water-treated control (Kuribara and Maruyama, 1996).

prolonged the time spent in the open arms without any significant change in the ambulatory activity. It is therefore expected that the plus-maze test in mice using the improved apparatus is adequate for evaluation of the anxiolytic effect of Kampo medicines.

4.3.2.3.1 *Anxiolytic effect of Saiboku-to fractions*

Since the seven daily treatments with Saiboku-to (0.5–2 g/kg) developed the anxiolytic effect in a dose-dependent manner, Kuribara *et al.* (1996) further tried to assess the anxiolytic potentials of the Saiboku-to fractions. Almost the same anxiolytic potential as that of Saiboku-to was observed when the chloroform-soluble fraction (F4, yield 1%), but not the water-insoluble (F1, yield 14%), water-soluble (F2, yield 30%) or ethanol-soluble (F3, yield 44%) fractions, was administered for 7 days. The chloroform-soluble fraction F4 was further divided into four subfractions (F4-1 to F4-4) according to the difference in the chloroform/methanol solubility by column chromatography (Maruyama *et al.*, 1998). Strong and mild anxiolytic effects were found in chloroform-soluble subfraction F4-1 (yield 11.5%) and the chloroform/methanol (30:1)-soluble subfraction F4-2 (yield 23.5%), respectively. Neither the chloroform/methanol (10:1)-soluble subfraction F4-3 (yield 50.9%) nor the chloroform/methanol (3:2)-soluble subfraction F4-4 (yield 5.5%) had such anxiolytic potential.

4.3.2.3.2 *Honokiol as the principal anxiolytic chemical in* Magnolia

GC-EIMS was then utilized to identify the active chemicals in Saiboku-to subfraction F4-1 (Maruyama *et al.*, 1998). From a search of the Wiley mass spectral library software, two peaks were found to match those of magnolol and honokiol, biphenyl compounds derived from *Magnolia* (Figure 4.3). Further confirmation of the identities of these biphenyl compounds was obtained by comparison with authentic magnolol and honokiol. As shown in Figure 4.4, the elevated plus-maze test in mice revealed that honokiol was the principal chemical for the anxiolytic effect of Saiboku-to. This

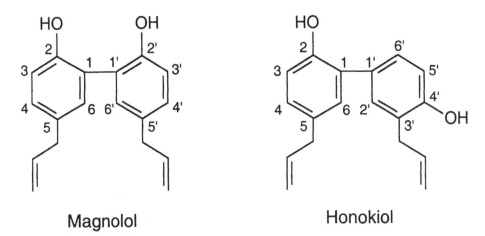

Figure 4.3 Chemical structures of magnolol and honokiol.

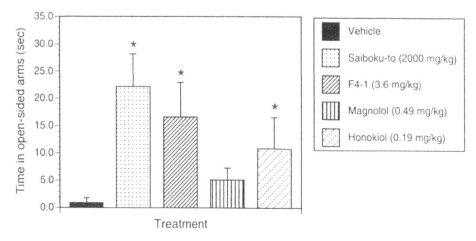

Figure 4.4 Anxiolytic potentials of Saiboku-to, chloroform-soluble Saiboku-to subfraction (F4-1), magnolol and honokiol. Each agent was administered daily for 7 days, and the elevated plus-maze test was carried out 24 h after the last administration. Values present the mean with SEM for 10 mice. * $p < 0.05$ compared to the vehicle-treatment group (Maruyama *et al.*, 1998).

was also supported by the time course of changes in the development of the anxiolytic effect following repeated administration. Similarly to the effect of authentic honokiol, a substantial anxiolytic effect was measured when either Saiboku-to or its subfraction F4-1 was administered for at least 5 days (Maruyama *et al.*, 1998). The effect reached a peak by the seventh administration, and almost the same anxiolytic potential was retained thereafter.

4.3.2.4 *Anxiolytic potentials of* Magnolia *and* Magnolia-*prescribed Kampo medicines: dependence on the honokiol content*

Although honokiol derived from *Magnolia* plays an important role in the development of the anxiolytic effect of Saiboku-to, there is a question whether the anxiolytic potential behaviorally determined is dependent on the content of honokiol in *Magnolia* and/or *Magnolia*-prescribed Kampo medicines.

4.3.2.4.1 Magnolia *samples*

Kuribara *et al.* (1999) compared the anxiolytic potentials of authentic honokiol and hot-water extracts of three *Magnolia* samples collected in different regions; two were Karakoboku (KA, from Zhejiang-sheng, China, honokiol 0.25% and magnolol 1.16%; and KB, from Sichuan-sheng, China, honokiol 1.72% and magnolol 1.71%), and one was Wakoboku (WA, from Iwate-ken, Japan, honokiol 0.32% and magnolol 0.81%). The doses of the extracts of the three *Magnolia* samples were adjusted so that the doses in terms of honokiol were 0.1 mg/kg (40 mg/kg KA), and 0.2 mg/kg (80 mg/kg KA, 11.6 mg/kg KB and 62.5 mg/kg WA).

In the elevated plus-maze test in mice, seven daily treatments with 40 and 80 mg/kg KA, which was the sample containing the lowest amount of honokiol (i.e., the dose of

Table 4.8 Effects of honokiol and magnolol, and the water extracts of three *Magnolia* samples assessed by elevated plus-maze and activity tests in mice

Drug	Dose	Plus-maze test	Activity test
		Time in open arms (s)	Counts/5 min
Vehicle		4.3 ± 1.4	29.5 ± 2.9
Honokiol	0.1 mg/kg	11.2 ± 2.8	32.2 ± 2.9
	0.2 mg/kg	17.2 ± 4.9**	28.5 ± 3.1
	0.5 mg/kg	11.8 ± 3.5	22.3 ± 1.3
	1.0 mg/kg	8.3 ± 1.9	26.3 ± 3.7
Magnolol	0.2 mg/kg	7.3 ± 3.4	27.8 ± 2.4
	1.0 mg/kg	10.2 ± 3.5	32.0 ± 2.8
Vehicle		5.3 ± 1.6	25.9 ± 2.2
KA	40.0 mg/kg (0.1 mg/kg)	13.0 ± 4.5	28.3 ± 3.6
	80.0 mg/kg (0.2 mg/kg)	16.1 ± 4.7*	29.6 ± 3.5
KB	11.6 mg/kg (0.2 mg/kg)	16.5 ± 4.0*	22.5 ± 2.3
WA	62.5 mg/kg (0.2 mg/kg)	16.2 ± 3.9*	22.2 ± 2.5

Vehicle: Aqueous solution containing 0.5% ethanol and Tween-80. KA: Karakoboku (*Magnoliae officinalis* from Zhejiang-sheng, China). KB: Karakoboku (*Magnoliae officinalis* from Sichuan-sheng, China). WA: Wakoboku (*Magnoliae obovate* from Iwate-ken, Japan).

Figures in parentheses for KA, KB and WA indicate the dose in terms of honokiol.
Each drug or Tween-80 (control) was orally administered daily for 7 days. The plus-maze test (5 min) was followed by the activity test (5 min), carried out 24 h after the last drug administration. The data presented are mean ± SEM ($n = 10$). * $p < 0.05$ and ** $p < 0.01$ compared to the vehicle-treated control (one-way ANOVA with Student's *t*-test) (Kuribara *et al.*, 1999).

magnolol administered was the highest), resulted in almost the same anxiolytic potentials as those of 0.1 and 0.2 mg/kg honokiol, respectively. The anxiolytic potentials of 11.6 mg/kg KB and 62.5 mg/kg WA were also similar to that of 0.2 mg/kg honokiol (Table 4.8). No significant change in the ambulatory activity was produced by any treatment with honokiol, KA, KB or WA. These results suggest that the anxiolytic potentials of the *Magnolia* extracts are simply dependent on the contents of honokiol.

4.3.2.4.2   *Hange-koboku-to and Saiboku-to samples*

Are the anxiolytic potentials of *Magnolia*-prescribed Kampo medicines also dependent on the content of honokiol? The evaluation of the anxiolytic potentials was carried out in five samples of each Hange-koboku-to or Saiboku-to which were prescribed with various *Magnolia*; two were Wakoboku (from Gifu-ken and Kouchi-ken, Japan), and three were Karakoboku (from Fujian-sheng, Sichuan-sheng and Yunnan-sheng, China) (Kuribara *et al.*, 2000a). The contents of honokiol and magnolol were different among the Kampo medicines: Hange-koboku-to (TJ-16A, honokiol 0.025% and magnolol 0.102%; TJ-16B, 0.085% and 0.053%; TJ-16C, 0.013% and 0.052%; TJ-16D, 0.106% and 0.083%; and TJ-16E, 0.138% and 0.137%); and Saiboku-to (TJ-96A, honokiol 0.008% and magnolol 0.028%; TJ-96B, 0.040% and 0.020%; TJ-96C, 0.007% and 0.025%; TJ-96D, 0.055% and 0.036%; and TJ-96E, 0.086% and 0.077%). The doses of test samples were also adjusted to ensure a constant dose of honokiol at 0.2 mg/kg, so that the doses of magnolol and undetermined chemicals varied among the test samples.

Table 4.9 Effects of honokiol, and Hange-koboku-to (TJ-16) and Saiboku-to (TJ-96), Kampo-medicines prescribed with various *Magnolia* barks, assessed by elevated plus-maze and activity tests in mice

Samples	Dose	n	Plus-maze test	Activity test
			Time in open arms (s) (% of honokiol)	Counts/5 min
Vehicle	10 ml/kg	50	7.6 ± 1.1	29.4 ± 1.4
Honokiol	0.2 mg/kg	50	22.8 ± 2.4* (100)	27.5 ± 1.3
TJ-16A	0.8000 g/kg	10	25.2 ± 10.0* (110.5)	20.4 ± 2.2
TJ-16B	0.2352 g/kg	20	20.4 ± 3.6* (89.5)	25.4 ± 2.1
TJ-16C	1.5384 g/kg	10	22.4 ± 5.4* (78.2)	27.9 ± 3.4
TJ-16D	0.1886 g/kg	10	22.7 ± 4.7* (99.6)	34.7 ± 3.5
TJ-16E	0.1450 g/kg	10	26.0 ± 6.2* (114.0)	28.9 ± 4.1
TJ-96A	2.5000 g/kg	19	22.3 ± 4.0* (97.8)	29.4 ± 2.7
TJ-96B	0.5000 g/kg	20	16.4 ± 4.0* (71.9)	24.5 ± 1.6
TJ-96C	2.8572 g/kg	20	10.8 ± 2.1* (47.4)	30.2 ± 2.2
TJ-96D	0.3636 g/kg	9	28.7 ± 5.3* (125.9)	30.9 ± 2.4
TJ-96E	0.2326 g/kg	10	25.5 ± 4.6* (111.8)	27.5 ± 2.9

Vehicle: Aqueous solution containing 0.5% ethanol and Tween-80. TJ-16A and TJ-16B, and TJ-96A and TJ-96B were prescribed with *Magnolia obovata* Thunb. (Japanese plants, Wakoboku), and the others with *Magnoliae officinalis* Rhad. et Wils (Chinese plants, Karakoboku). The dose of each sample was adjusted to ensure the constant dose of honokiol at 0.2 mg/kg. Vehicle, honokiol and test samples were orally administered daily for 7 days. The plus-maze test (5 min) was carried out 24 h after the last drug administration, and was subsequently followed by the activity test (5 min). The data presented are mean ± SEM. * $p < 0.05$ compared to the group treated with vehicle (Student–Newman–Keuls test) (Kuribara *et al.*, 2000a).

The seven daily treatments with 9 out of 10 test samples developed anxiolytic potentials as high as that caused by 0.2 mg/kg honokiol (Table 4.9). The only exception was the sample of Saiboku-to (TJ-96C) that contained the lowest amount of honokiol, meaning that comparatively higher doses of undetermined chemical(s) were administered in this test sample. The *Magnolia*-free samples of Hange-koboku-to or Saiboku-to never developed the anxiolytic effect.

The results obtained from the two experiments using *Magnolia* samples and *Magnolia*-prescribed Kampo medicines suggest that honokiol is the principal chemical responsible for the anxiolytic effect of water extract of *Magnolia* and the two Kampo medicines Hange-koboku-to and Saiboku-to, and that the chemicals derived from plants other than *Magnolia* are responsible for mildly influencing the anxiolytic effect of honokiol. It is also expected that the elevated plus-maze test in mice is applicable for estimation of the content of honokiol in *Magnolia* as well as in *Magnolia*-prescribed Kampo medicines in addition to the assessment of their anxiolytic potential.

4.3.2.5 Characteristics of the anxiolytic effect of honokiol

Kuribara *et al.* (1998) behaviorally evaluated in detail the characteristics and mechanisms of the anxiolytic effect of honokiol, and compared them with those of the typical benzodiazepine anxiolytic diazepam.

4.3.2.5.1   Single administration of honokiol

In the plus-maze test using the BALB/c mice, a significant prolongation of the time spent in the open arms (i.e., anxiolytic effect) was manifested 3 h after the single oral administration of 20 mg/kg honokiol. However, the effect disappeared within 24 h after administration. Honokiol at this dose did not change the ambulatory activity or the traction performance.

4.3.2.5.2   Seven-day administration of honokiol

The seven daily treatments with 0.2 mg/kg honokiol significantly prolonged the time spent in the open arms when the plus-maze test was carried out 3 h after the last administration. At 24 h after the last administration, doses of honokiol 0.1–0.5 mg/kg caused prolongation in the time spent in the open arms in a dose-dependent manner. Following 1 and 2 mg/kg honokiol, however, the anxiolytic potentials were slightly lower than that following 0.5 mg/kg. Neither ambulatory activity nor traction performance was affected by any treatment with honokiol (Table 4.10).

These results indicate that the repeated administration of honokiol causes an enhancement of the anxiolytic potential to approximately 100 times that following the single administration. It is therefore probable that an accumulation of active chemical(s), generated from the metabolism of honokiol, is responsible for the development of the anxiolytic effect.

4.3.2.5.3   Combined drug administrations

The results following the combined administrations of honokiol or diazepam with various drugs (Kuribara *et al.*, 1998) are presented in Table 4.11. Combination of honokiol and diazepam significantly enhanced the anxiolytic effect. In contrast, the diazepam-induced increase in the ambulatory activity, but not disruption of the traction performance, was ameliorated by honokiol. Flumazenil (0.3 mg/kg s.c.) reduced

Table 4.10 Effects of seven daily oral treatments with honokiol assessed by plus-maze and activity tests in mice

Treatments	n	Plus-maze test	Activity test
		Time in open arms (s)	Counts/5 min
3 h after the last administration			
Tween-80	10	10.2 ± 3.7	30.9 ± 2.3
Honokiol 0.2 mg/kg	10	49.8 ± 10.8*	31.8 ± 3.9
24 h after the last administration			
Tween-80	10	14.5 ± 5.2	30.3 ± 2.9
Honokiol 0.1 mg/kg	10	26.8 ± 8.0	25.6 ± 3.8
0.2 mg/kg	10	43.0 ± 7.0*	27.1 ± 2.3
0.5 mg/kg	10	58.6 ± 13.2*	27.9 ± 3.8
1.0 mg/kg	10	41.2 ± 7.7*	29.9 ± 2.0
2.0 mg/kg	10	45.8 ± 6.8*	36.8 ± 2.6

* $p < 0.05$ compared to the Tween-80-treated control group (Kuribara *et al.*, 1998).

Table 4.11 Effects of oral honokiol (0.2 mg/kg, seven-day treatment) or diazepam (1 mg/kg), either alone or in combination with other drugs, assessed by plus-maze, activity and traction tests in mice

Treatments	Plus-maze test	Activity test	Traction test
	Time in open arms (s)	Counts/5 min	Clinging time (s)
Tween-80	12.3 ± 2.6	24.9 ± 1.7	60.0 ± 0.0
Honokiol 0.2 mg/kg	$43.0 \pm 7.0*$	27.1 ± 2.3	60.0 ± 0.0
Diazepam 1.0 mg/kg	$43.5 \pm 6.1*$	$38.9 \pm 3.2*$	$42.6 \pm 6.3*$
Honokiol + diazepam	$102.3 \pm 13.6\#$	$29.8 \pm 4.4\#$	$43.8 \pm 4.9*$
Flumazenil 0.3 mg/kg	6.7 ± 2.9	25.6 ± 3.0	60.0 ± 0.0
Honokiol + flumazenil	$9.7 \pm 5.1\#$	25.5 ± 2.3	60.0 ± 0.0
Diazepam + flumazenil	$13.5 \pm 3.8\#$	$27.9 \pm 1.8\#$	$54.5 \pm 2.9*\#$
Bicuculline 0.1 mg/kg	7.1 ± 2.8	$21.5 \pm 2.7*$	60.0 ± 0.0
Honokiol + bicuculline	$10.7 \pm 4.7\#$	$20.9 \pm 1.4*\#$	60.0 ± 0.0
Diazepam + bicuculline	$17.4 \pm 4.9\#$	$37.0 \pm 2.3*$	$46.6 \pm 6.9*$
CCK-4 50 μg/kg	$2.2 \pm 1.1*$	29.2 ± 2.2	60.0 ± 0.0
Honokiol + CCK-4	$9.3 \pm 3.2\#$	33.8 ± 3.0	60.0 ± 0.0
Diazepam + CCK-4	$57.3 \pm 13.0*$	34.1 ± 5.1	$41.3 \pm 7.7*$
Caffeine 30 mg/kg	13.9 ± 9.0	$60.5 \pm 7.4*$	60.0 ± 0.0
Honokiol + caffeine	$31.0 \pm 8.0*$	$63.7 \pm 4.4*\#$	60.0 ± 0.0
Diazepam + caffeine	$93.8 \pm 11.3*\#$	$55.1 \pm 7.1*\#$	50.4 ± 5.1

In the honokiol study, 24 h after the last administration of honokiol, bicuculline, CCK-4 or caffeine was administered and the behavioral tests were conducted 10, 10, 10 or 15 min, respectively, after administration. In the diazepam study, diazepam, bicuculline and CCK-4 were administered 10 min before, and caffeine 15 min before the behavioral tests. Values are mean $\pm$ SEM, $n = 10$. * $p < 0.05$ compared to the Tween-80-treated group. # $p < 0.05$ compared to the honokiol- or diazepam-treated group (Kuribara *et al.*, 1998).

the honokiol- and diazepam-induced prolongation in the time spent in the open arms. Similar drug interactions were also demonstrated following the combined treatment with *Magnolia*-prescribed Kampo medicines Hange-koboku-to and Saiboku-to (Kuribara and Maruyama, 1995, 1996), and Saiboku-to extracts (Kuribara *et al.*, 1996). Flumazenil also ameliorated the diazepam-induced increase in the ambulatory activity and disruption of the traction performance.

The anxiolytic effect of honokiol was inhibited by the $GABA_A$ receptor antagonist bicuculline (0.1 mg/kg s.c.) and the anxiogenic agents cholecystokinin tetrapeptide (CCK-4, 50 μg/kg i.p.) and caffeine (30 mg/kg i.p.). The anxiolytic effect of diazepam (1 mg/kg p.o.) was inhibited by flumazenil and bicuculline. Diazepam completely reversed the anxiogenic effect of CCK-4. However, the combined treatment with diazepam and caffeine caused further prolongation in the time spent in the open arms.

Taken together, these results suggest that, in contrast to the benzodiazepine anxiolytic diazepam, honokiol is likely to possess selective and potent anxiolytic effects without eliciting the behavioral disorders that are frequently produced as side effects of benzodiazepine anxiolytics, such as sedation/disinhibition, muscle relaxation, ataxia and motor impairment (Schweizer *et al.*, 1995; Woods and Winger, 1995; Woods *et al.*, 1992, 1995). Two distinct mechanisms are possible. Honokiol or its metabolites may selectively stimulate the $GABA_A$/benzodiazepine receptor subtypes that are responsible for the anxiolytic effect, or they may bind the sites responsible for the development of the anxiolytic effect other than the $GABA_A$/benzodiazepine receptors.

4.3.2.6   *Anxiolytic effect of dihydrohonokiol (DHH-B)*

While quite a high dose of honokiol, 20 mg/kg, is required to develop a significant anxiolytic effect after single administration, 0.2 mg/kg honokiol is sufficient for development of the anxiolysis following seven daily administrations (Kuribara *et al.*, 1998; Maruyama *et al.*, 1998). The half-lives of honokiol after i.v. administration of 5 and 10 mg/kg honokiol were 49.22 ± 6.68 min and 56.25 ± 7.30 min, respectively (Tsai *et al.*, 1995), indicating that it is difficult to consider accumulation of honokiol as the mechanism of the enhancement of the anxiolytic effect after the repeated administration. Rather, the disagreement between the pharmacokinetic and pharmacodynamic results indicates the possibility that accumulation of metabolite(s) of honokiol is responsible for the development of the anxiolytic effect.

The repeated administration of magnolol yielded tetrahydromagnolol (Hattori *et al.*, 1984, 1986), indicating that metabolism of honokiol may also yield similar hydrogenated derivatives, and that they are involved in the development of the anxiolytic effect. If such a hypothesis is true, the hydrogenated derivative(s) of honokiol can show anxiolytic effects even after a single administration. To confirm this hypothesis, the anxiolytic potentials were evaluated following a single oral administration of honokiol and its eight analogues (Figure 4.5) (Kuribara *et al.*, 2000b).

4.3.2.6.1   *Assessment by the plus-maze test*

Among the eight analogues of honokiol, one partially reduced derivative, dihydrohonokiol (DHH-B; 3'-(2-propenyl)-5-propyl-(1,1'-biphenyl)-2,4'-diol)), exhibited a significant anxiolytic effect even at 0.04 mg/kg. Following oral administration of 1 mg/kg DHH-B, the anxiolytic effect was evident at 2 h, peaked at 3 h, and remained at a significant level for longer than 4 h; for i.p. administration the respective times were 1 h, 2 h and 3 h. No derivatives including DHH-A, except for DHH-B3 being mildly anxiolytic, exhibited significant anxiolytic effect (Tables 4.12 and 4.13). No marked change in the motor activity was produced by either derivative of honokiol. These results strongly suggest that the reduction of one of two side chains of the propenyl group to a propyl group is important for development of the anxiolytic effect.

The combined administration of DHH-B and diazepam enhanced the anxiolytic effect. However, the anxiolytic effect of DHH-B was not significantly antagonized by flumazenil (0.3 mg/kg) or bicuculline (0.1 mg/kg). These results imply that DHH-B develops its anxiolytic effect through an action other than $GABA_A$/benzodiazepine receptor complexes.

4.3.2.6.2   *Assessment by the Vogel conflict test*

The Vogel conflict test in rats, originally established by Vogel *et al.* (1971), is a simple and reliable method for evaluation of anxiolytic effect of drugs. To reconfirm and emphasize the anxiolytic effct of DHH-B, the conflict test was carried out in mice

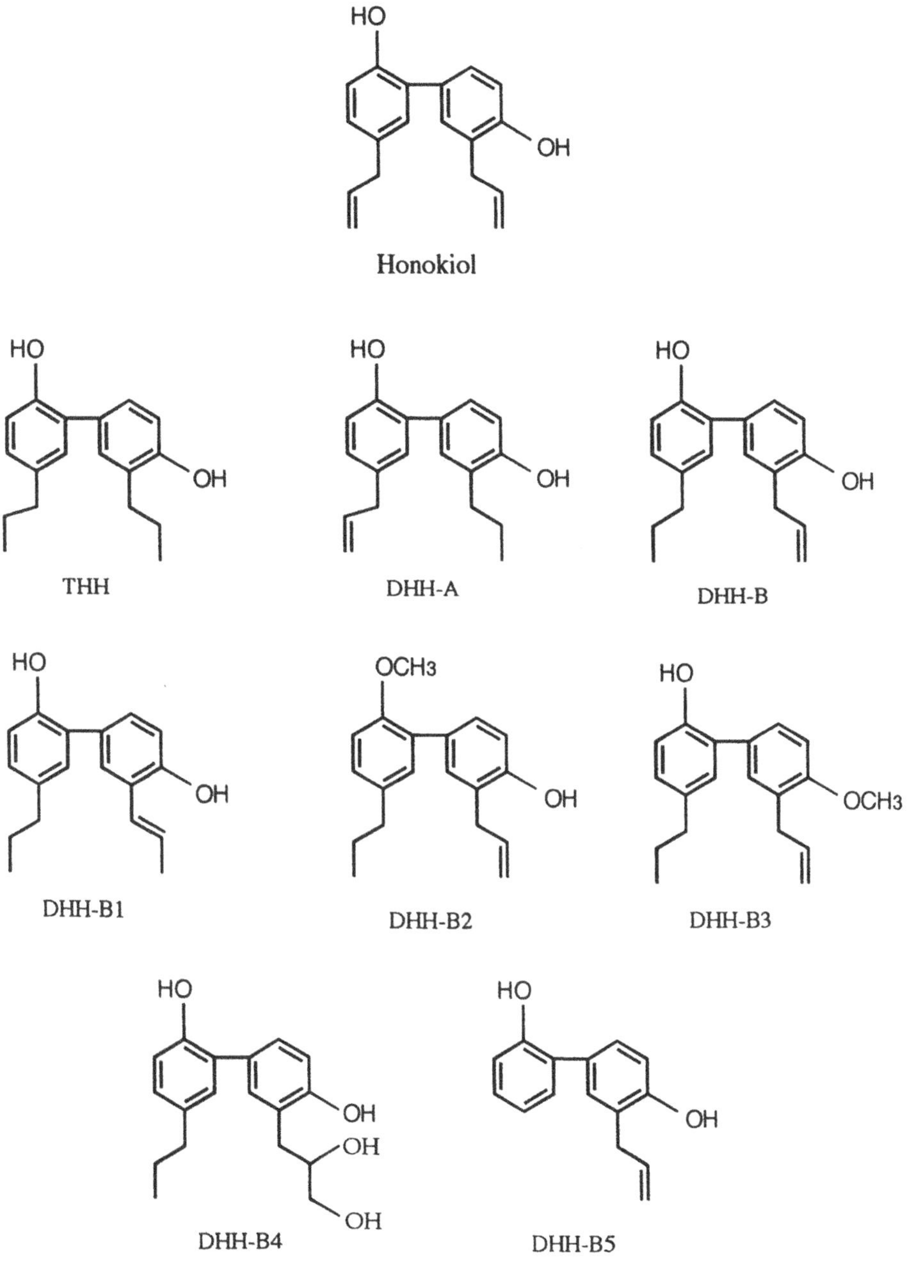

Figure 4.5 Chemical structures of honokiol and its eight analogues.

(Kuribara *et al.*, 1989; Umezu, 1999), by a slight modification of the original Vogel method.

Briefly, mice that had been deprived of water for 24 h were individually placed into the chambers, and allowed a free intake of water from the spout for 30 min. They were then kept in their home cages without a water supply. On the next day the mice were

Table 4.12 Effects of single oral administration of honokiol, dihydrohonokiol (DHH-A and DHH-B) and tetrahydrohonokiol (THH), and diazepam assessed by elevated plus-maze and activity tests in mice

Treatments		Plus-maze test	Activity test
		Time in open-arms (s)	Counts/5 min
Vehicle		5.8 ± 2.0	32.4 ± 1.7
Honokiol	5.0 mg/kg	6.5 ± 4.2	30.6 ± 2.7
	10.0 mg/kg	5.1 ± 2.2	35.0 ± 2.1
	20.0 mg/kg	8.7 ± 2.3	40.4 ± 4.2
DHH-A	0.04 mg/kg	1.0 ± 0.8	28.9 ± 3.4
	0.2 mg/kg	5.4 ± 2.9	22.4 ± 2.7
	1.0 mg/kg	10.1 ± 3.9	35.3 ± 4.7
DHH-B	0.008 mg/kg	6.8 ± 2.0	35.9 ± 3.0
	0.04 mg/kg	12.3 ± 3.4*	36.4 ± 2.1
	0.2 mg/kg	18.1 ± 5.6*	33.0 ± 2.4
	1.0 mg/kg	21.5 ± 5.2*	33.0 ± 2.4
THH	1.0 mg/kg	5.2 ± 1.7	31.1 ± 2.9
	10.0 mg/kg	5.3 ± 2.3	31.7 ± 1.8
Vehicle		3.2 ± 1.3	28.6 ± 3.0
Diazepam	1.0 mg/kg	22.1 ± 5.8*	33.6 ± 3.1

Honokiol, DHH-A, DHH-B and THH were administered 3 h before, and diazepam 10 min before the behavioral tests. * $p < 0.05$ compared to the vehicle-treated control mice. $n = 10$ in each group except for the vehicle-treated control ($n = 20$) (Kuribara *et al.*, 2000b).

Table 4.13 Effects of single oral administration of derivatives of honokiol (B1 through B6, 2 mg/kg) assessed by elevated plus-maze and activity tests in mice

Treatments	Plus-maze test	Activity test
	Time in open-arms (s)	Counts/5 min
1 h after administration		
Vehicle	3.9 ± 1.4	29.3 ± 3.6
B1	4.3 ± 2.0	42.0 ± 3.6*
B2	8.7 ± 2.1	37.6 ± 3.5
B3	6.3 ± 2.2	37.4 ± 4.2
B4	14.8 ± 5.3*	39.2 ± 3.8*
B5	8.7 ± 2.2	33.6 ± 2.9
3 h after administration		
Vehicle	1.3 ± 0.8	31.0 ± 3.2
B1	6.1 ± 2.3	40.8 ± 3.8
B2	5.3 ± 2.7	36.9 ± 3.3
B3	9.6 ± 3.8*	39.1 ± 4.9
B4	2.9 ± 1.8	33.6 ± 4.1
B5	1.3 ± 0.7	36.6 ± 4.6

* $p < 0.05$ compared to the Tween-80-treated control; $n = 10$ in each group (Kuribara *et al.*, 2000b).

Table 4.14 Anxiolytic potentials of dihydrohonokiol (DHH-B) and diazepam evaluated by two different behavioral tests in mice

Drug	Dose (mg/kg)	n	Plus-maze test		Vogel's conflict test	
			Time in open arm (s)	n	Punished drinkings/30 min	
Vehicle		20	5.8 ± 2.0	30	4.3 ± 1.5	
DHH-B	0.008	10	6.8 ± 2.0			
	0.04	10	12.3 ± 3.4*			
	0.2	10	18.1 ± 5.6*			
	1.0	10	21.5 ± 5.2*	20	7.1 ± 3.7	
	2.0			10	6.0 ± 2.8	
	5.0			10	17.9 ± 5.6**	
Vehicle		10	3.2 ± 1.3	30	4.3 ± 1.5	
Diazepam	1.0	10	22.1 ± 5.8*	30	12.8 ± 2.7*	

Each figure represents mean ± SEM * $p < 0.05$, ** $p < 0.01$, compared to the vehicle-treated control (Kuribara *et al.*, unpublished data).

returned to the same chambers. During this stage, namely the conflict session, the mice could drink water for 30 min under condition of suffering a punishment of an electric foot-shock (45 V, 0.2 mA, 50 Hz AC, for 0.3 s), delivered through the stainless-steel floor grid of the chamber, with each water consumption of 0.05 ml. Drugs were administered prior to the conflict session. A significant increase in the punished drinking was produced by the administration of DHH-B (5 mg/kg p.o., 3 h before). Diazepam (1 mg/kg p.o., 10 min before) also increased the punished drinking (Table 4.14).

These results strongly support the finding of the elevated plus-maze test that DHH-B is a potent anxiolytic compound, and that this compound is the principal active metabolite of honokiol.

4.3.3   Conclusion

Magnolia-prescribed Kampo medicines have been empirically used to relieve anxiety, nervous tension, and/or insomnia. An improved elevated plus-maze test in mice was applied to assess the anxiolytic effect. The seven-day treatment with either Hange-koboku-to or Saiboku-to prolonged the time spent in the open arms (i.e., anxiolytic effect), and the anxiolytic potentials were roughly dependent on the content of honokiol derived from *Magnolia* bark. The anxiolytic potential of honokiol following seven-day treatment was approximately 100 times that following the single treatment. Furthermore, a single administration of dihydrohonokiol (DHH-B) exhibited an anxiolytic effect 400 times as potent as that of honokiol. The combined administration of honokiol or DHH-B with diazepam, a benzodiazepine anxiolytic, enhanced the anxiolytic effect. The anxiolytic effects of these chemicals were antagonized by flumazenil, a benzodiazepine antagonist.

Taken together, these results indicate that honokiol derived from *Magnolia* is the main chemical responsible for the anxiolytic effect of *Magnolia*-prescribed Kampo medicines, and that the hydrogenation of honokiol to DHH-B is an important pathway in enhancing the anxiolytic effect following repeated administration of honokiol.

References for section III

Dawson, G.R. and Tricklebank, M.D. (1995) Use of the elevated plus-maze in the search for novel anxiolytic agents. *Trends Pharmacol. Sci.*, 16, 33–36.

Hattori, M., Sakamoto, T., Endo, Y., Kakiuchi, N., Kobashi, K., Mizuno, T. and Namba, T. (1984) Matabolism of magnolol from magnoliae cortex. I. Application of liquid chromatography-mass spectrometry to the analysis of metabolites of magnolia in rats. *Chem. Pharm. Bull.*, 32, 5010–5017.

Hattori, M., Endo, Y., Takebe, S., Kobashi, K., Fukusaku, N. and Namba, T. (1986) Metabolism of magnolol from magnoliae cortex. II. Absorption, metabolism and excretion of [*ring*-^{14}C]magnolol in rats. *Chem. Pharm. Bull.*, 34, 158–167.

Hosoya, E. and Yamamura, Y. (eds) (1988) *Recent Advances in the Pharmacology of Kampo (Japanese Herbal) Medicines.* Excerpta Medica International Congress Series 85. Amsterdam: Excerpta Medica.

Ishida, H., Ootake, T., Kuribara, H. and Maruyama, Y. (1999) Clinical study of the augmentative effect of Saiboku-to for anxiolytic and antidepressive action of diazepam. *Pain Clin.*, 20, 395–399. [In Japanese]

Juangsu New Medical College (1985) *Zhong Yao Da Ci Dian (Dictionary of Chinese Materia Medica)*, pp. 1628–1639. Shanghai: Shanghai Scientific and Technological Publishers.

Kuribara, H. and Maruyama, Y. (1995) Assessment of anxiolytic effect of Hange-koboku-to with elevated plus-maze test in mice. *Shinkei Seishin Yakuri*, 17, 353–358. [Abstract in English]

Kuribara, H. and Maruyama, Y. (1996) The anxiolytic effect of oriental herbal medicines by an improved plus-maze test in mice: involvement of benzodiazepine receptors. *Shinkei Seishin Yakuri*, 18, 179–190. [Abstract in English]

Kuribara, H., Haraguchi, H. and Tadokoro, S. (1989) Anticonflict effect of caffeine: investigation by punishment and hypertonic NaCl solution procedures in mice. *Jpn. J. Alcohol Drug Depend.*, 24, 144–153.

Kuribara, H., Morita, M., Ishige, A., Hayashi, K. and Maruyama, Y. (1996) Investigation of the anxiolytic effect of the extracts derived from Saiboku-to, an oriental herbal medicine, by an improved plus-maze test in mice. *Shinkei Seishin Yakuri*, 18, 643–653. [Abstract in English]

Kuribara, H., Stavinoha, W.B. and Maruyama, Y. (1998) Behavioral pharmacological characteristics of honokiol, an anxiolytic agent present in extracts of magnolia bark, evaluated by an elevated plus-maze test in mice. *J. Pharm. Pharmacol.*, 50, 819–826.

Kuribara, H., Kishi, E., Hattori, N., Yuzurihara, M. and Maruyama, Y. (1999) Application of the elevated plus-maze test in mice for evaluation of the content of honokiol in water extracts of magnolia. *Phytother. Res.*, 13, 593–596.

Kuribara, H., Kishi, E., Hattori, N., Okada, M. and Maruyama, Y. (2000a) The anxiolytic effect of two Kampo medicines in Japan prescribed with magnolia bark is derived from honokiol. *J. Pharm. Pharmacol.*, 52, 1425–1429.

Kuribara, H., Kishi, E., Kimura, M., Weintraub, S.T. and Maruyama, Y. (2000b) Comparative assessment of the anxiolytic activities of honokiol and derivatives. *Pharmacol. Biochem. Behav.*, 67, 597–601.

Lister, R.G. (1987) The use of plus-maze to measure anxiety in the mouse. *Psychopharmacology*, 92, 180–185.

Maruyama, Y., Kuribara, H., Morita, M., Yuzurihara, M. and Weintraub, S.T. (1998) Identification of magnolol and honokiol as anxiolytic agents in extracts of Saiboku-to, an oriental herbal medicine. *J. Nat. Prod.*, 61, 135–138.

Namba, T. and Tsuda, Y. (eds) (1998) *Outline of Pharmacognosy, a Textbook*, pp. 395–422. Tokyo: Nankodo [In Japanese]

Narita, H. (1990) Use of Kampo medicines in psychiatry. *Shinkei Seishin Yakuri*, 12, 165–172. [In Japanese]

Nishimoto, K. (1986) The quality of magnolia bark. *Gendai Toyo Igaku*, 7, 68–72. [In Japanese]

Nishioka, I. (1983) Pharmacology of magnolia bark. *Kampo Igaku*, 7, 1–2. [In Japanese]

Pellow, S., Chopin, P., File, S.E. and Briley, M. (1985) Validation of open:closed arm entries in an elevated plus-maze as a measure of anxiety in the rat. *J. Neurosci. Methods*, 14, 149–169.

Schweizer, E., Rickels, K. and Uhlenhuth, E.H. (1995) Issues in the long-term treatment of anxiety disorders. In *Psychopharmacology: The Fourth Generation of Progress*, edited by F.E. Bloom and D.J. Kupfer, pp. 1349–1359. New York: Raven Press.

Tsai, T.H., Chou, C.J. and Chen, C.F. (1995) Disposition of magnolol after intravenous bolus and infusion in rabbits. *Drug Metab. Dispos.*, 22, 518–521.

Umezu, T. (1999) Effects of psychoactive drugs in the Vogel conflict test in mice. *Jpn. J. Pharmacol.*, 80, 111–118.

Vogel, J.R., Beer, B. and Clody, D.E. (1971) A simple and reliable conflict procedure for testing antianxiety agents. *Psychopharmacologia (Berlin)*, 21, 1–7.

Watanabe, K. (1994) Pharmacology of magnolia bark and its constituents. *Gendai Toyo Igaku*, 15, 419–424.

Woods, J.H. and Winger, G. (1995) Current benzodiazepine issues. *Psychopharmacology*, 118, 107–115.

Woods, J.H., Katz, J.L. and Winger, G. (1992) Benzodiazepines: use, abuse, and consequences. *Pharmacol. Rev.*, 44, 151–347.

Woods, J.H., Katz, J.L. and Winger, G. (1995) Abuse and therapeutic use of benzodiazepines and benzodiazepine-like drugs. In *Psychopharmacology: The Fourth Generation of Progress*, edited by F.E. Bloom and D.J. Kupfer, pp. 1777–1791. New York: Raven Press.

SECTION IV PHARMACOLOGICAL CHARACTERISTICS OF *MAGNOLIA* EXTRACTS: MAGNOLOL AND HONOKIOL

Hisashi Kuribara

4.4.1 Introduction

In the preceding section, it was shown in terms of the results from an improved elevated plus-maze test in mice that the anxiolytic effect of *Magnolia* and *Magnolia*-prescribed Kampo medicines is derived from dihydrohonokiol (DHH-B), a biphenyl compound generated from honokiol. A role of the $GABA_A$/benzodiazepine receptor complex is suggested in the development of the anxiolytic effect of DHH-B.

As demonstrated following the administration of diazepam, benzodiazepine anxiolytics sometimes develop side effects through inhibiting the central nervous system functions as well as tolerance and dependence (Schweizer *et al.*, 1995). In this section, the pharmacological characteristics of the *Magnolia* extracts honokiol and its isomer magnolol are reviewed to emphasize the comparatively lower risk of benzodiazepine-like side effects from these preparations and chemicals compared to those from diazepam. In addition, a possible process of generation of DHH-B from honokiol is presented.

4.4.2 Central depressive effects of *Magnolia* extracts

4.4.2.1 *Muscle relaxant effect*

Watanabe *et al.* (1973) reported the central depressive effect of extracts of *Magnolia*. After intraperitoneal administration of the water extract (1 g/kg), mice developed muscle weakness in 2–3 min, followed by cessation of breathing and heart beat in a few minutes. In contrast, no marked behavioral symptoms, except for a slight decrease in spontaneous motor activity for a short period, were produced by oral administration of the water extract (2 g/kg). In the case of the ether extract, however, both oral and intraperitoneal administrations resulted in strong muscle relaxation and disruption of clinging to a steel net for up to 90–120 min, without death. The ED_{50} values for the muscle relaxant effect of intraperitoneally administered ether extract of *Magnolia* and its alkaline soluble fraction were estimated to be 582 mg and 580 mg, respectively (Watanabe *et al.*, 1975). The ethanol extract showed an intermediate effect between those of the water and ether extracts. These results indicate that the strong lethal effect of intraperitoneally injected water extract is caused by the peripherally mediated curare-like muscle relaxant action of magnocurarine and its related compounds (Tomita *et al.*, 1951), and the ether extract shows a central depressive action. The known chemicals in the ether extract are essential oils such as pinenes, camphene, limonene, and α-, β- and γ-eudesmol (Fujita *et al.*, 1973a; Itokawa, 1986) and phenolic compounds such as magnolol, honokiol and magnaldehyde (Fujita *et al.*, 1972, 1973b,c; Yamazaki, 1986; Yahara *et al.*, 1991).

4.4.2.2   *Anticonvulsive effect*

The central depressive effect of the ether extract of *Magnolia* is also supported by its anticonvulsive effect. The intraperitoneal pretreatment of mice with the ether extract (1 mg/kg, 50 min before) and its alkali-soluble fraction completely protected the convulsions and death caused by strychinine (1 mg/kg s.c.) and picrotoxin (7.5 mg/kg s.c.), and the tremor induced by oxotremorine (1 mg/kg i.p.). The same pretreatment with the ether extract antagonized the pentetrazol (120 mg/kg s.c.)-induced extensor tonus and death, but not the convulsive twitch.

The ether extract of *Magnolia* and the alkali-soluble fractions inhibited the spinal reflexes, the crossed extensor reflex responding to the stimulation of the central end of the cut sciatic nerve, in young chicks (ED_{50} 48.8 mg/kg and 21.2 mg/kg, respectively) (Watanabe *et al.*, 1973). This effect was antagonized by strychnine (0.5 mg/kg i.p.). The tonic extensor convulsion produced by intracerebroventricular injection of penicillin G potassium (50 μg) in mice was also inhibited by *Magnolia* ether extract (ED_{50} 530 mg/kg) and its alkali-soluble fraction (ED_{50} 251 mg/kg).

It was noted that the anticonvulsive potentials of the ether extract were approximately half those of the alkali-soluble fraction, suggesting that multiple compounds having different chemical characteristics are responsible for the anticonvulsive effect of the ether extract.

4.4.3   Central depressive effects of magnolol and honokiol

Watanabe *et al.* (1975, 1983a,b) evaluated in detail the central depressive effects of magnolol and honokiol. Both intraperitoneally administered magnolol and honokiol disrupted the grip strength in mice for up to 180 min (ED_{50} 131 mg/kg and 217 mg/kg, respectively) (Watanabe *et al.*, 1975), and inhibited the spinal reflexes in chicks for up to 60 min (ED_{50} 10.3 mg/kg and 11.1 mg/kg, respectively) (Watanabe *et al.*, 1983a). Magnolol also inhibited penicillin G potassium-induced convulsions in mice (ED_{50} 31.3 mg/kg) (Watanabe *et al.*, 1983b).

However, it should be considered that considerably higher doses of magnolol and honokiol were administered to develop such behavioral symptoms. Since the concentration of honokiol in *Magnolia* bark is 0.25–1.7% (Kuribara *et al.*, 1999a), the doses of magnolol and honokiol effective for the development of central depression and muscle relaxation correspond to 15–156 g/kg of *Magnolia*. In contrast, the elevated plus-maze test in mice revealed that seven daily treatments with honokiol and extracts of *Magnolia* developed the anxiolytic effect even at 0.2 mg/kg (Kuribara *et al.*, 1998) and 11.6–80 mg/kg (Kuribara *et al.*, 1999a, 2000a), respectively, as did single treatment with 20 mg/kg honokiol (Kuribara *et al.*, 1998). Thus, assessment of pharmacological effects of honokiol at doses equal to the effective doses for anxiolytic effect is proposed.

4.4.4 Mechanisms of the central depressive effects of magnolol and honokiol

Electrophysiological experiments were carried out to evaluate the mechanism of the central depressive effect of magnolol and honokiol. Magnolol (30 mg/kg) caused a decrease in spontaneous arousal pattern in the hippocampus and induced periodic spindle bursts in the sensory and motor cortices (Watanabe *et al.*, 1983a). Comparatively higher concentrations of magnolol (10^{-5} and 10^{-4} M) selectively inhibited the size of spontaneous ventral root potential without eliciting a marked change in the dorsal root potential, and also inhibited the ventral root potential induced by glutamate and aspartate (Kudo and Watanabe, 1984). However, the GABA-induced potential could not be inhibited by magnolol. These findings indicate that the muscle relaxing effect of high doses of magnolol is produced via inhibition of the stimulatory neurotransmission and increase in the threshold of firing of motor neurons.

Neurochemical evaluation revealed that honokiol (1–10 μM), but not magnolol (1–10 μM), elicited a concentration-dependent enhancement of 25 mM K^+-evoked acetylcholine (ACh) release from rat hippocampus slices (Tsai *et al.*, 1995b). Addition of tetrodotoxin (1 μM), pilocarpine (1 μM) or methoctamine (1 μM) had no effect on the enhanced ACh release by honokiol. These results indicate that part of the central depressive effect of honokiol is produced through an enhancement of K^+-evoked ACh release directly on hippocampal cholinergic terminals via receptors other than M_2 cholinergic subtypes. Taken together, the results from the behavioral, electrophysiological and neurochemical evaluations strongly suggest that magnolol and honokiol show an inhibitory effect on several areas of the brain such as the hypothalamic and reticular formation ascending activating systems as well as the spinal cord.

4.4.5 Interaction with centrally acting drugs

An evaluation of the interactions of Saiboku-to, one of the Kampo medicines prescribed with *Magnolia*, and its fractions with the centrally acting drugs methamphetamine and haloperidol was conducted using a discrete shuttle avoidance response in mice (Aoki *et al.*, 1997). Methamphetamine (0.5 mg/kg s.c.) stimulated the avoidance response, causing a marked increase in the response rate, whereas haloperidol (0.1 mg/kg s.c.) significantly suppressed the avoidance response, decreasing both the response rate and the percent avoidance.

A single oral administration of 2 g/kg Saiboku-to mildly decreased the response rate and the percent avoidance, suppressing the avoidance response. The haloperidol-induced suppression of the avoidance response was enhanced by single treatment with Saiboku-to (2 g/kg p.o.), and its ethanol and chloroform extracts. However, methamphetamine-induced avoidance stimulation was not affected by any treatment.

A preliminary investigation of the interaction of honokiol with centrally acting drugs was also carried out. Honokiol (0.1–10 mg/kg p.o.) did not change the avoidance response following single administration, and nonsignificantly modified the methamphetamine-induced avoidance stimulation or haloperidol-induced avoidance suppression (Kuribara *et al.*, unpublished data).

These results indicate that the interactions of honokiol at the doses for anxiolytic effect, and probably of magnolol, with centrally acting drugs are almost negligible and that the central depressive effect of Saiboku-to demonstrated in the shuttle avoidance test is caused by unknown chemical(s) rather than by honokiol. The characteristics of the effect of DHH-B on the discrete shuttle avoidance response in mice were almost the same as those of honokiol (Kuribara *et al.*, unpublished data).

4.4.6 Benzodiazepine-like side effects of *Magnolia*-prescribed Kampo medicines

Kiwaki *et al.* (1989) evaluated side effects following oral treatment with Saiboku-to (0.5–2 g/kg) for 90 days in rats. All treated rats survived and showed no abnormal signs in terms of body weight, food consumption or gross behaviors as well as the pathological or biochemical indicators.

In our experiments for evaluation of the anxiolytic effect of Kampo medicines, 7 to 14 daily administrations of 0.5–2 g/kg of either Hange-koboku-to or Saiboku-to, or Saiboku-to extracts did not develop any behavioral signs related to benzodiazepine-like side effects.

4.4.7 Benzodiazepine-like side effects of honokiol

Benzodiazepine anxiolytics frequently cause central depressive symptoms such as ataxia, oversedation, amnesia, ethanol and barbiturate potentiation, and tolerance and dependence even at therapeutic doses. Such unwanted side effects limit the clinical usefulness of benzodiazepine anxiolytics (Schweizer *et al.*, 1995; Woods *et al.*, 1992, 1995; Woods and Winger, 1995). Kuribara *et al.* (1999b) evaluated behaviorally the side effects of honokiol at doses required for anxiolytic effect, and the effects were compared to those of diazepam.

4.4.7.1 *Muscle relaxation*

The traction test in mice (Kuribara *et al.*, 1977) was conducted to evaluate the muscle relaxant effect of drugs. Briefly, a wire (diameter 1.6 mm and length 30 cm) was set horizontally at a height of 30 cm. The mouse was first forced to grasp the wire with the four paws, and the duration of clinging to the wire was measured for up to 60 s. When the duration of clinging was >60 s the mouse was released from the wire and the clinging time was recorded as 60 s.

Diazepam at 0.5–2 mg/kg, doses that were equivalent to the anxiolytic doses, disrupted the traction performance in a dose-dependent manner. Neither single treatment with 2–20 mg/kg honokiol nor seven daily treatments with 0.1–2 mg/kg honokiol caused significant change in the traction performance. These results indicate that, in contrast to diazepam, honokiol has less liability to producte muscle relaxation after acute or chronic administration at the doses that produce the anxiolytic effect.

4.4.7.2 Physical dependence

To assess whether honokiol induced benzodiazepine-like physical dependence, mice were orally treated with honokiol (0.1–2 mg/kg) or diazepam (1–10 mg/kg) daily for 12 days following the method originally described by Cumin *et al.* (1982) with slight modifications. Subsequently, all mice were challenged intraperitoneally with flumazenil (10 mg/kg) 24 h after the last drug treatment, and the occurrences of behaviors indicating withdrawal symptoms—hyperreactivity (vocalization induced by a light pushing of the back), tremor, clonic convulsion, tonic convulsion, tail-flick/reaction, and running fits (wild running evoked by a key-ring sound)—were observed for 30 min. During the measurement of tremor, clonic and tonic convulsions, and tail-flick/reaction, no external stimulation was presented to the mouse.

As shown in Table 4.15, the mice treated with diazepam 1 mg/kg and higher doses showed precipitated withdrawal symptoms characterized by hyperreactivity, tremor and running fits when they were challenged with flumazenil. Although some mice

Table 4.15 Number of mice exhibiting symptoms following challenge with flumazenil (10 mg/kg i.p.)

Pretreatment (12 days)	Challenge	HR	TR	CC	TC	TF	RF
Tween-80	Saline	1/10	0/10	0/10	0/10	0/10	0/10
Tween-80	Flumazenil	3/10	0/10	0/10	0/10	1/10	0/10
Honokiol (mg/kg)							
0.1	Flumazenil	4/10	0/10	0/10	0/10	0/10	0/10
0.2	Flumazenil	4/10	0/10	0/10	0/10	0/10	0/10
0.5	Flumazenil	4/10	0/10	0/10	0/10	0/10	0/10
1.0	Flumazenil	5/10	0/10	0/10	0/10	0/10	0/10
2.0	Flumazenil	5/10	0/10	0/10	0/10	0/10	0/10
Diazepam (mg/kg)							
0.5	Flumazenil	7/10	0/10	0/10	0/10	0/10	0/10
1.0	Flumazenil	10/10*	0/10	0/10	0/10	0/10	3/10
2.0	Flumazenil	10/10*	5/10*	0/10	0/10	2/10	5/10*
5.0	Flumazenil	10/10*	7/10*	0/10	0/10	1/10	8/10*
10.0	Flumazenil	10/10*	5/10*	3/10	1/10	1/10	10/10*

Tween-80, honokiol and diazepam were administered p.o. once a day for 12 days, and the challenge with flumazenil was carried out 24 h after the last treatment. The figures presented are the numbers of mice that showed the symptoms during the observation period for 30 min.
HR, hyper-reactivity (vocalization induced by a light pushing of the back); TR, tremor; CC, clonic convulsion; TC, tonic convulsion; TF, tail flick or tail reaction; RF, running fits (wild running auditorily evoked by key-ring sound).

* $p < 0.05$ compared to mice treated with Tween-80 and challenged with flumazenil (Kuribara *et al.*, 1999b).

Table 4.16 Effects of honokiol (p.o. treatment 3 h before) and diazepam (p.o. treatment 10 min before) on the hexobarbital (100 mg/kg i.p.)-induced sleeping

	Sleeping time (s)
Tween-80	2552 ± 209
Honokiol (mg/kg)	
0.2	2454 ± 254
2.0	2334 ± 129
20.0	2480 ± 12
Tween-80	2902 ± 84
Diazepam (mg/kg)	
0.5	3688 ± 91*
1.0	4476 ± 336*
2.0	5370 ± 480*
Honokiol (0.2 mg/kg) + diazepam (1 mg/kg)	4144 ± 457
Honokiol (2.0 mg/kg) + diazepam (1 mg/kg)	4466 ± 232
Honokiol (20.0 mg/kg) + diazepam (1 mg/kg)	5040 ± 472

* $p < 0.05$ compared to Tween-80-treated control; $n = 5$ in each group (Kuribara *et al.*, 1999b).

treated with 10 mg/kg diazepam suffered tonic and clonic convulsions after challenge with flumazenil, all the mice survived. Challenge with flumazenil was not followed by any behavioral signs, except for a mild hyperreactivity, in the mice treated with 0.1–2 mg/kg honokiol. These results indicate that honokiol has very low ability, if any, for induction of benzodiaepine-like physical dependence even at doses 10 times higher than the minimum anxiolytic dose.

4.4.7.3   *Barbiturate potentiation*

To evaluate the centrally depressive effects of drugs, mice were treated orally with honokiol (0.2–20 mg/kg) or diazepam (0.5–2 mg/kg). Subsequently, they were given hexobarbital (100 mg/kg i.p.), and the latency times of loss and recovery of righting reflex were recorded. The difference was considered as the sleeping time.

As shown in Table 4.16, diazepam dose-dependently prolonged the hexobarbital-induced sleeping time, reflecting the centrally depressive effect. However, honokiol did not modify either the hexobarbital-induced sleeping time or the enhancement by diazepam of the hexobarbital-induced sleeping time. These findings indicate that, as compared to the anxiolytic effect, the centrally depressive effect of honokiol is very mild, and that honokiol has less ability for induction of strong central depression even after combined administration with drugs having a centrally depressive effect such as barbiturates, ethanol, benzodiazepines, etc.

4.4.7.4   *Amnesia*

To evaluate whether honokiol caused amnesia, a learning and memory test in mice using an elevated plus-maze (Itoh *et al.*, 1990, 1991) was carried out with slight modifications. Briefly, in the training trial (first day) each mouse was placed at the end of one of the two open arms, which was randomly selected, facing away from the

Table 4.17 Effects of honokiol (p.o. treatment 3 h before) and diazepam (p.o. treatment 10 min before) on the transfer latency in the training and retention trials

	Training trial	Retention trial
Administration before the training trial		
Tween-80	59.1 ± 9.8	25.6 ± 3.3
Honokiol (mg/kg)		
2	60.3 ± 6.9	39.0 ± 4.8*
20	50.6 ± 7.5	18.1 ± 3.0
Tween-80	59.1 ± 10.1	32.5 ± 9.5
Diazepam (mg/kg)		
0.5	27.2 ± 6.1*	60.5 ± 13.6*
1.0	41.8 ± 9.3	109.2 ± 16.9*
2.0	69.6 ± 17.9	114.3 ± 22.8*
Administration before the retention trial		
Tween-80	62.8 ± 12.2	25.2 ± 2.8
Honokiol (mg/kg)		
2	51.2 ± 10.3	31.9 ± 4.0
20	59.7 ± 7.4	26.8 ± 6.3
Tween-80	71.7 ± 10.4	31.9 ± 9.9
Diazepam (mg/kg)		
0.5	60.8 ± 18.9	24.2 ± 3.6
1.0	85.3 ± 13.2	37.1 ± 5.8
2.0	63.9 ± 15.7	30.0 ± 3.7

Each figure represents the transfer latency (s) ± SEM from the open arm to the closed arm.
* $p < 0.05$ compared to Tween-80-treated control (Kuribara *et al.*, 1999b).

central platform. The latency time of transfer from the open arm to either of the closed arms (the transfer latency) was recorded, and the mouse was allowed to move freely in the plus-maze for 2 min. Then the mouse was gently returned to its home cage. On the next day, the retention trial was carried out. The mouse was again placed in the same position as in the training trial, and the transfer latency was recorded. Honokiol (2–20 mg/kg) or diazepam (0.5–2 mg/kg) was administered orally before either the training or the retention trial.

As shown in Table 4.17, treatment with honokiol before either the training or the retention trial did not change the transfer latencies in either trials, except for a mild prolongation in the transfer latency in the retention trial by the administration of 2 mg/kg honokiol before the training trial. However, there was no significant difference between control and honokiol (2 mg/kg)-treated groups in the gross times of shortening of the transfer latencies in the training and retention trials. In contrast, pretraining administration of diazepam prolonged the latency in the retention trial in a dose-dependent manner. No significant change in the latency was produced when diazepam was administered before the retention trial. These results suggest that, consistent with clinical evidence that the amnesic effect of benzodiazepine anxiolytics is mediated by the inhibition of the input of information into the storage site rather than the disruption of memory recall (Haefely *et al.*, 1990; Doble and Martin, 1992), diazepam carries a risk of amnesic effect at the anxiolytic doses, while it is indicated that honokiol is much less likely than diazepam to inducte amnesia at anxiolytic doses.

4.4.8 Side effects of DHH-B

Essentially, DHH-B demonstrated almost the same characteristics as those of hono-kiol (Kuribara *et al.*, 2000c). Thus, neither acute nor chronic treatment with DHH-B developed diazepam-like side effects.

4.4.9 Pharmacokinetics of magnolol and honokiol

Pharmacokinetic studies revealed that the half-lives of disposition in the rat plasma samples were 54.1 ± 5.14, 49.05 ± 5.96 and 49.58 ± 6.81 min after i.v. administration of magnolol at 2, 5 and 10 mg/kg, respectively (Tsai *et al.*, 1995a), and 49.22 ± 6.68 and 56.2 ± 7.30 min after honokiol at 5 and 10 mg/kg, respectively (Tsai *et al.*, 1994). These results indicate that magnolol and honokiol show similar pharmacokinetic characteristics, and their half-lives of disposition are almost independent of the doses; that is, the kinetics of disposition are first-order.

There were species differences in the kinetics of disposition of magnolol between rats and rabbits (Tsai *et al.*, 1995a). Thus, the half-lives of magnolol in rabbit (14.5 ± 1.77 min for 5 mg/kg i.v. bolus administration, and 15.7 ± 3.00 min for 76 µg/kg/min i.v. infusion) were approximately one-third of those in rats. However, the results in rabbits also support the consideration that the pharmacokinetics of disposition of magnolol, and probably of honokiol, are first-order independently of the administration routes, doses and animal species.

4.4.10 Metabolism of magnolol and honokiol

After i.v. administration, magnolol distributed almost uniformly in the rat brain tissues (Tsai *et al.*, 1996). Magnolol was metabolized to isomagnolol (the propenyl side chains being transformed to allyl groups), hydrogenated and hydroxy derivatives, glucuronides and sulfates (Tsai *et al.*, 1995c, 1996). Furthermore, it was suggested that tissue enzymes and intestinal bacterial enzymes were involved in the metabolism of orally administered magnolol. The distribution in the brain and metabolism of honokiol may be similar to those of magnolol.

Hattori *et al.* (1984, 1986) determined the metabolites of magnolol after repeated oral administration to rats using high-performance liquid chromatography and liquid chromatography–mass spectrometry. The urinary and fecal metabolites of magnolol were tetrahydromagnolol (M1), 5-(1-propen-1(E)-yl)-5'-propyl-2,2'-dihydroxybiphenyl (M2), 5-allyl-5'-propyl-2,2'-dihydroxybiphenyl (M3), isomagnolol (5,5'-di(1-propen-1(E)-yl)-2,2'-dihydroxybiphenyl) (M4), and 5-allyl-5'-(1-propen-1(E)-yl)-2,2'-dihydroxybiphenyl (M5). In the feces, in 24 h after the first administration of 50 mg/kg magnolol only a small amount of metabolites were detected, but magnolol was recorded as a major constituent (22% of the administered dose). Isomagnolol and tetrahydromagnolol, however, increased in amount almost linearly during 48–72 h accompanied by a significant decrease in the amount of magnolol. The former metabolite became constant in amount after 96 h, but the latter increased further and finally reached a maximum level after 120 h. Because the pharmacokinetic characteristic of honokiol is similar to that of magnolol (Tsai *et al.*, 1994, 1996), it is expected that repeated administration of honokiol will result in a change in the generation of metabolites of honokiol.

The elution patterns of urinary metabolites of orally administered magnolol were essentially similar to those observed in the case of the fecal metabolites. However, the *in vitro* incubation of magnolol with rat bacteria did not yield tetrahydromagnolol, suggesting that the intestinal bacteria under anaerobic conditions mostly take part in the isomerization of magnolol but not in the reduction (hydrogenation).

Hattori *et al.* (1986) evaluated the metabolism of [*ring*-^{14}C]magnolol after single oral administration. There were two peaks of the blood radioactivity level at 15 min and 8 h, suggesting an enterohepatic circulation of magnolol and its metabolites. A major metabolite excreted in the bile was [*ring*-^{14}C]magnolol-2-0-glucuronide. After oral and intraperitoneal administration of [*ring*-^{14}C]magnolol, most of radioactivity was eliminated into feces and urine within the first 12 h in each case.

4.4.11   Conclusion

The central effects of *Magnolia* extracts as well as honokiol and magnolol have been reviewed. The ether extract of *Magnolia* as well as honokiol and magnolol caused centrally mediated muscle relaxation and depression via inhibition of several areas of the brain such as the hypothalamic and reticular formation ascending activating systems as well as the spinal cord. However, at the effective doses for the anxiolytic effect, neither honokiol nor dihydrohonokiol (DHH-B) produced any symptoms such as ataxia, muscle relaxation, oversedation, amnesia, ethanol and barbiturate potentiation, tolerance and dependence, or interaction with centrally acting drugs. These results suggest that honokiol and DHH-B develop their anxiolytic effect with low risk of benzodiazepine-like side effects.

Pharmacokinetic studies revealed that the half-lives of honokiol and magnolol in rat blood samples were almost the same for single i.v. administration and for continuous i.v. infusion, although there were differences in the half-lives between rat and rabbit. Furthermore, it was suggested that intestinal bacteria are not involved in the hydrogenation of orally administered honokiol and magnolol.

References for section IV

Aoki, A., Kuribara, H. and Maruyama, Y. (1997) Assessment of psychotropic effects of Saiboku-to by discrete shuttle avoidance in mice. *Shinkei Seishin Yakuri*, 19, 357–364. [Abstract in English]

Cumin, R., Bonett, E.P., Scherschlicht, R. and Haefely, W.E. (1982) Use of specific benzodiazepine antagonist, Ro-15-1788, in studies of physiological dependence on benzodiazepines. *Experientia*, 38, 833–834.

Doble, A. and Martin, I.L. (1992) Multiple benzodiazepine receptors: no reason for anxiety. *Trends Pharmacol. Sci.*, 13, 76–81.

Fujita, M., Itokawa, H. and Sashida, Y. (1972) Honokiol, a new phenolic compound isolated from the bark of *Magnolia obovata* THUNB. *Chem. Pharm. Bull.*, 20, 212–213.

Fujita, M., Itokawa, H. and Sashida, Y. (1973a) Studies of the compounds of *Magnolia obovata* THUNB. I. On the components of the essential oil of the bark. *Yakugaku Zasshi*, 93, 415–421. [Abstract in English]

Fujita, M., Itokawa, H. and Sashida, Y. (1973b) Studies of the compounds of *Magnolia obovata* THUNB. II. On the components of the methanol extract of the bark. *Yakugaku Zasshi*, 93, 422–428. [Abstract in English]

Fujita, M., Itokawa, H. and Sashida, Y. (1973c) Studies of the compounds of *Magnolia obovata* THUNB. III. Occurrence of magnolol and honokiol in *Magnolia obovata* and other allied plants. *Yakugaku Zasshi*, 93, 429–434. [Abstract in English]

Haefely, W.E., Martin, J.R. and Schoch, P. (1990) Novel anxiolytics that act as partial agonists at benzodiazepine receptors. *Trends Pharmacol. Sci.*, 11, 452–456.

Hattori, M., Sakamoto, T., Endo, Y., Kakiuchi, N., Kobashi, K., Mizuno, T. and Namba, T. (1984) Metabolism of magnolol from magnoliae cortex. I. Application of liquid chromatography–mass spectrometry to the analysis of metabolites of magnolia in rats. *Chem. Pharm. Bull.*, 32, 5010–5017.

Hattori, M., Endo, Y., Takebe, S., Kobashi, K., Fukusaku, N. and Namba, T. (1986) Metabolism of magnolol from magnoliae cortex. II. Absorption, metabolism and excretion of [ring-^{14}C]magnolol in rats. *Chem. Pharm. Bull.*, 34, 158–167.

Itoh, J., Nabeshima, T. and Kameyama, T. (1990) Utility of an elevated plus-maze for the evaluation of memory in mice: effects of nootropics, scopolamine and electroconvulsive shock. *Psychopharmacology*, 101, 27–33.

Itoh, J., Nabeshima, T. and Kameyama, T. (1991) Utility of an elevated plus-maze for dissociation of amnesic and behavioral effects of drugs in mice. *Eur. J. Pharmacol.*, 194, 71–76.

Itokawa, H. (1986) Chemistry of the *Magnolia obovata* Thunberg. *Gendai Toyo Igaku*, 7, 60–67. [In Japanese]

Kiwaki, S., Yamashita, Y., Yamamae, H., Nakamura, T. and Oketani, Y. (1989) Ninety-day toxicity study of a Chinese herb medicine, Saiboku-to extract in rats. *Oyo Yakuri*, 38, 495–509. [Abstract in English]

Kudo, Y. and Watanabe, K. (1984) Inhibitory effects of magnolol, an effective compound of magnolia bark, on amino acid-induced root potentials in frog spinal cord. *Wakan Iyakugaku Zasshi*, 1, 108–109 [In Japanese].

Kuribara, H., Higuchi, Y. and Tadokoro, S. (1977) Effects of central depressants on rota-rod and traction performances in mice. *Jpn. J. Pharmacol.*, 27, 117–126.

Kuribara, H., Stavinoha, W.B. and Maruyama, Y. (1998) Behavioral pharmacological characteristics of honokiol, an anxiolytic agent present in extracts of magnolia bark, evaluated by an elevated plus-maze test in mice. *J. Pharm. Pharmacol.*, 50, 819–826.

Kuribara, H., Kishi, E., Hattori, N., Yuzurihara, M. and Maruyama, Y. (1999a) Application of the elevated plus-maze test in mice for evaluation of the content of honokiol in water extracts of magnolia. *Phytother. Res.*, 13, 593–596.

Kuribara, H., Stavinoha, W.B. and Maruyama, Y. (1999b) Honokiol, a putative anxiolytic agent extracted from magnolia bark, has no diazepam-like side effects in mice. *J. Pharm. Pharmacol.*, 51, 97–103.

Kuribara, H., Kishi, E., Hattori, N., Okada, M. and Maruyama, Y. (2000a) The anxiolytic effect of two Kampo medicines in Japan prescribed with magnolia bark is derived from honokiol. *J. Pharm. Pharmacol.*, 52, 1425–1429.

Kuribara, H., Kishi, E. and Maruyama, Y. (2000c) Does dihydrohonokiol, a potent anxiolytic compound, develop benzodiazepine-like side effects? *J. Pharm. Pharmacol.*, 52, 1017–1022.

Schweizer, E., Rickels, K. and Uhlenhuth, E.H. (1995) Issues in the long-term treatment of anxiety disorders. In *Psychopharmacology: The Fourth Generation of Progress*, edited by F.E. Bloom and D.J. Kupfer, pp. 1349–1359. New York: Raven Press.

Tomita, M., Inubuchi, Y. and Yamada, M. (1951) Studies of the alkaloids of magnoliaceous plants I. Alkaloids of *Magnolia obovata* THUNB. *Yakugaku Zasshi*, 71, 1069–1075. [Abstract in English]

Tsai, T.H., Chou, C.J., Cheng, F.C. and Chen, C.F. (1994) Pharmacokinetics of honokiol after intravenous administration in rats assessed using high-performance liquid chromatography. *J. Chromatogr. Biomed. Appl.*, 655, 41–45.

Tsai, T.H., Chou, C.J. and Chen, C.F. (1995a) Disposition of magnolol after intravenous bolus and infusion in rabbits. *Drug Metab. Dispos.*, 22, 518–521.

Tsai, T.H., Chou, C.J. and Chen, C.F. (1995b) Glucuronidation of magnolol assessed using HPLC/fluorescence. *Planta Med.*, 61, 491–492.

Tsai, T.H., Westly, J., Lee, T.F., Chen, C.F. and Wang, L.C.H. (1995c) Effects of honokiol and magnolol on acetylcholine release from rat hippocampal slices. *Planta Med.*, 61, 477–479.

Tsai, T.H., Chou, C.J. and Chen, C.F. (1996) Pharmacokinetics and brain distribution of *Magnolol* in the rat after intravenous bolus injection. *J. Pharm. Pharmacol.*, 48, 57–59.

Watanabe, K., Goto, Y. and Yoshitomi, K. (1973) Central depressant effects of the extracts of magnolia cortex. *Chem. Pharm. Bull.*, 21, 1700–1798.

Watanabe, K., Watanabe, H., Goto, Y., Yamamoto, N. and Yoshizaki, M. (1975) Studies on the active principles of magnolia bark. Centrally acting muscle relaxant activity of magnolol and honokiol. *Jpn. J. Pharmacol.*, 25, 605–607.

Watanabe, K., Watanabe, H., Goto, Y., Yamaguchi, M., Yamamoto, N. and Hagino, K. (1983a) Pharmacological properties of magnolol and honokiol extracted from *Magnolia officinalis*: central depressant effects. *Planta Med.*, 49, 103–108.

Watanabe, H., Watanabe, K. and Hagino, K. (1983b) Chemostructural requirement for centrally acting muscle relaxant effect of magnolol and honokiol, neolignane derivatives. *J. Pharm. Dyn.*, 6, 184–190.

Wood, J.H. and Winger, G. (1995) Current benzodiazepine issues. *Psychopharmacology*, 118, 107–115.

Woods, J.H., Katz, J.L. and Winger, G. (1992) Benzodiazepines: use, abuse, and consequences. *Pharmacol. Rev.*, 44, 151–347.

Woods, J.H., Katz, K.L. and Winger, G. (1995) Abuse and therapeutic use of benzodiazepines and benzodiazepine-like drugs. In *Psychopharmacology: The Fourth Generation of Progress*, edited by F.E. Bloom and D.J. Kupfer, pp. 1777–1791. New York: Raven Press.

Yahara, S., Nishiyori, T., Kohda, A., Nohara, T. and Nishioka, I. (1991) Isolation and characterization of phenolic compounds from magnoliae cortex produced in China. *Chem. Pharm. Bull.*, 39, 2024–2036.

Yamazaki, M. (1986) The prescription of Magnoliae cortex in the traditional Chinese medicine. *Gendai Toyo Igaku*, 7, 47–53.

5   Quality Control of *Magnolia* Bark

Peter X. Zhang, Amanda Harris and Yu Shao

5.1   Introduction

The bark of *Magnolia* has been used in traditional medicinal preparations in China, Japan, and Korea to treat many ailments ranging from bronchitis and emphysema to neurosis to gastrointestinal disorders, and its clinical effects have been proven (Zheng *et al.*, 1999). The *Magnolia* bark used as a plant drug is collected from trees of over 16 years old and the raw material in China has historically been in short supply (Su *et al.*, 1992). The Chinese Pharmacopoeia (1995) lists the dried bark of *Magnolia officinalis* Rehd. et Wils. and *M. officinalis* var. *biloba* Rehd. et Wils. as genuine species. However, owing to the shortage of raw material, there are at least 40 species of plants from nine families that have been used as genuine species or alternatives to the genuine species by local herbal healers (Su *et al.*, 1992; Zheng *et al.*, 1999). Some of them showed similar therapeutic effects to the genuine species, but most did not (Zheng *et al.*, 1999).

Traditional Chinese medicine (TCM) uses herbal formulas that combine several herbs together. Herbal formulas have been prescribed to deliver the maximum and/or balanced effects of plant drugs, but it has also made the identification of ingredients and evaluation of the ingredients' pharmacological effects more complicated. Furthermore, with modern extraction techniques, there is a tendency to make plant drugs or herbal formulas into extracts and eventually process them into pharmaceutical forms, such as granules, capsules, tablets, and injection solutions. Extraction concentrates the active components, but it changes the physical appearances of plant drugs. Therefore, reliable testing methods for identification of plant drugs and quantitative analysis of their marker constituents are essential for quality assurance (QA) or quality control (QC) of plant drugs, including *Magnolia* bark and its preparations.

Quality control of *Magnolia* bark and its preparation deals not only with identification of *Magnolia* bark and determination of the concentration of its marker constituents, it also includes the testing of physical characteristics and the presence of contamination, such as pesticides, heavy metals, residual solvent, and microbial impurities. Product stability, safety, packaging, and storage also fall under the heading of quality control. Table 5.1 shows the tentative product specification of *Magnolia* bark powdered extract, which may help to give the reader an idea of the broad range of quality control.

Laboratory tests for QA/QC are roughly divided into two categories: physical tests and chemical tests. Physical tests include loss on drying, ash content, bulk density,

Table 5.1 Nutratech product specification of *Magnolia* bark

Product name:	*Magnolia* bark powdered extract
Botanical name:	*Magnolia officinalis*
Part(s) used:	Cortex
Solvent used:	70% ethanol
Carrier(s) used:	None
Assay	
Appearance:	Dark brown fine powder
Odor:	Weak, aromatic
Taste:	Neutral and characteristic
Bulk density:	< 0.56 g/ml untapped; > 0.72 g/ml tapped
Solubility:	Freely dissolves in 70% ethanol
pH:	2–4 (10% water solution)
Identification:	TLC confirm with reference standard
Quantitation:	HPLC: magnolol plus honokiol > 15%; honokiol > 4.5%
Heavy metals:	< 10 ppm
Pesticides residues:	< 10 ppb
Residual solvent:	< 0.05%
Microbial purity:	
Total plate count:	< 1000 cfu/g[a]
Yeast and molds:	< 100 cfu/g
Salmonella:	Absent
Escherichia coli:	Absent
Stability:	Under assay
Storage:	Cool, dry place
MSDM:[b]	Available upon request

[a] cfu, colony-forming unit(s).
[b] Material Safety Data Sheet.

microscopy, as well as comparison of the appearance, color, odor, and taste of the botanical material with a known reference standard. Physical tests are quick and simple, but it might require years of experience to acquire enough knowledge to do them accurately. For example, microscopic techniques for plant drug identification require experience if one is to know the histology of the genuine drug and its common adulterants. Owing to the variation of physical appearances from batch to batch and the changes in microscopic structures of plant drugs during processing, there are limits to the application of physical testing methods for the identification of processed materials, such as powdered extract, tincture, and formulated herbal tablets or capsules. The difficulty of documenting the physical tests also limits their application as key methods for quality control. Total ash of *Magnolia* bark and *Magnolia* bark powder should be no more than 6.0% (w/w). This is the only available quantitative physical test data for *Magnolia* barks (Japanese Pharmacopoeia, 1996).

Chemical tests, on the other hand, focus on determining the nature of the constituents in plant drugs (fingerprints) and determining the concentration of marker constituents. These tests are routinely used for QA/QC of plant drugs. Plant drugs have varying amounts of chemical constituents. Standardizing plant drugs and their preparations to a certain amount of the marker constituent(s) has been accepted as good manufacturing practice (GMP) in the production of phyto-medical and nutriceutical products.

In this chapter we review some published chemical methods for qualitative and quantitative analysis of *Magnolia* bark and its preparations. Then we present an HPLC method development and its validation in detail, with emphasis on chemical analysis as a process for QA/QC of plant drugs.

5.2   Chemical analysis as a process for quality control

Chemical analysis as a process for quality control of plant drugs consists of six steps: (i) choosing unique or biologically active marker compounds; (ii) obtaining a representative sample; (iii) choosing an appropriate analytical method; (iv) separating the analyte from any interfering substance; (v) performing a quantitative measurement; and (vi) evaluating the results.

5.2.1   *Choosing unique or biologically active marker compounds*

The chemical constituents of *Magnolia* bark have been well studied (Zhang, 1989; Chen *et al.*, 1997; Zheng *et al.*, 1999) and classified as follows.

- **Phenols:** Magnolol; honokiol; isomagnolol; tetrahydromagnolol; bornylmagnolol; piperitylmagnolol; piperibylholokiol; dipiperitylnolol; magnatriol; magnaldehyde B, C, D, E; magnilognan A, B, C, D, E, F, G, H, I; randainal; randaiol; syringaresinol; 6-0-methylhonokiol, and others.
- **Essential oils:** About 1% of the raw bark. More than 20 essential oils have been identified, and the major constituents are β-eudesmol (machilol), α-eudesmol, α-pinene, β-pinene, camphene, limonene, bornyl acetate, caryophyllene epoxide, and cryptomeridiol.
- **Alkaloids:** Magnocurarine, magnoflorine, anonaine, michelarbine, liriodenine, salicifoline, tubocurarine, and others.
- **Other:** Sinapicaldehyde, among others.

None of the known constituents is unique to *Magnolia* bark, and therefore they cannot be used as a fingerprint marker for identification. However, pharmacological studies have demonstrated that magnolol, honokiol, magnocuraine, and β-eudesmol are the major active components in *Magnolia* bark (Maruyama *et al.*, 1998; Zheng *et al.*, 1999; Wang *et al.*, 1999). Although magnocuraine and β-eudesmol have been used as markers, magnolol and honokiol are the most appropriate marker compounds for identification and quality control of *Magnolia* bark (Su *et al.*, 1992; Ye *et al.*, 1992). The Chinese Pharmacopoeia (1995) and the Japanese Pharmacopoeia (1996) have accepted magnolol and honokiol as marker constituents for QA/QC of *Magnolia* bark. Establishing and validating the chemical testing methods for identification and quantitative analysis of magnolol and honokiol are analytical problems to be solved for QA/QC of *Magnolia* bark.

5.2.2   *Obtaining a representative sample*

Obtaining a representative sample is essential for chemical tests of plant drugs. It sounds simple, but many useless results have been generated due to careless sampling or lack of knowledge of the characteristic of plant drugs. Fujita *et al.* (1973) studied the seasonal and age-related variation of magnolol and honokiol in *Magnolia obovata*

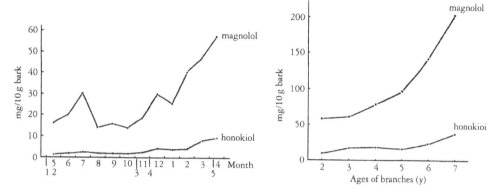

Figure 5.1 (a) Seasonal variation of magnolol and honokiol contents (mg in 10 g dried bark) in branches of *Magnolia obovata* in their first year. 1, Leaves unfold; 2, comes into flower; 3, leaves turn brown; 4, leaves shed; 5, puts forth leaves. (b) Age-related variation of magnolol and honokiol contents (mg in 10 g dried bark) in branches of *Magnolia obovata*.

Thunb. They sampled the first-year branch barks from May to April in the middle of each month, and sampled 2- to 7-year-old branch barks in the middle of April from the same tree. They then compared the content of magnolol and honokiol in those samples using thin-layer chromatography scanning and gas–liquid chromatography methods. Their results, shown in Figure 5.1, are valuable because their sampling procedure is well designed.

Ye *et al.* (1992) compared the quality of *Magnolia officinalis* and its allied plants from different areas in China using β-eudesmol, magonolol, and honokiol as markers. Their analytical work was excellent, but their conclusion regarding the geographic variation of *Magnolia* bark quality is questionable. Since the samples tested were purchased from traditional Chinese medicine pharmacies, no information about the age of the trees, harvest seasons, preparation after harvest, and storage conditions of the samples was taken into consideration.

Obtaining an authenticated plant drug as a reference standard is very important for quality control of plant drugs. Since the marker components are not unique to one species of plant, the identification of plant drugs by comparing their fingerprints under identical experimental conditions with an authenticated plant drug is common in pharmacognosy.

5.2.3 *Choosing an appropriate analytical method*

The routine chemical tests for QA/QC of plant drugs include, but are not limited to: total extractives; thin-layer chromatography (TLC); ultraviolet (UV)/visible (vis) spectrophotometry; near infrared (IR) or Fourier transform infrared (FTIR) spectroscopy; gas chromatography (GC); and high performance liquid chromatography (HPLC).

Chromatography is a group of techniques used to separate components of a mixture on an adsorbent column in a flowing system. This technique is the most widely used for QA/QC of plant drugs. Figure 5.2 shows the classification of chromatographic methods according to the types of mobile phase and stationary phases.

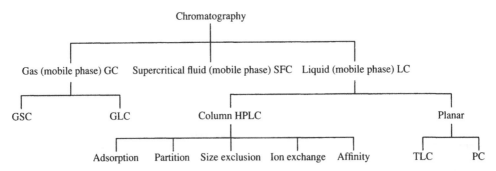

Figure 5.2 Classification of chromatographic methods according to mobile and stationary phases and their abbreviations: GC, gas chromatography; GSC, gas–solid chromatography; GLC, gas–liquid chromatography; LC, liquid chromatography; HPLC, high performance liquid chromatography; TLC, thin-layer chromatography; PC, paper chromatography; SFC, supercritical fluid chromatography.

TLC: TLC has been employed as the principal means of identification of plant drugs and plant drug preparations. The advantages of TLC are that (i) the time required for the demonstration of most of the constituents of a plant drug is very short; (ii) it provides semiquantitative information on major constituents, and a chromatographic fingerprint; (iii) it can identify plant drug(s) in drug combination; (iv) it requires less equipment, laboratory space and chemicals; (v) it is relatively cheap and easy to perform. The disadvantages of TLC are inaccuracy in quantitative analysis and difficulty of long-term data storage.

GC: Gas chromatography (GC) provides fast and accurate separation of volatile organic compounds. GC is not suitable for thermolabile compounds.

HPLC: HPLC has become the method of choice for a vast array of analytical separations because of its high resolution, selectivity, sensitivity, and speed. HPLC separation is carried out at or near room temperature, which is most suitable for analysis of natural products, including *Magnolia* bark. HPLC cannot be used for volatile compounds, and uses large amounts of organic solvents.

SFC: Supercritical fluid chromatography uses a supercritical fluid as the mobile phase. Supercritical carbon dioxide is the most commonly used mobile phase. Since the mobile phase evaporates instantly at the collecting tube(s), SFC has proven to be a powerful method for solid phase separation and has been used for isolation of plant constituents. SFC is an environmentally friendly method and is gaining popularity in research laboratories.

CE: Capillary electrophoresis separation is based on mobility differences of analytes in an electric field. CE uses small amounts of a sample, and needs no organic solvents for separation. With the growing environmental concerns over waste disposal and the high cost of biological samples, interest in CE is increasing. Zhang *et al.* (1997) published a capillary zone electrophresis (CZE) method for the separation and determination of magnolol and honokiol in *Magnolia officinalis* bark. CZE is the simplest and most widely used mode of CE. The authors suggested that CZE could be a potential alternative to HPLC for analysis of *Magnolia* bark. However, the troubleshooting in CE is not as well defined as it is in HPLC, which limits its popularity as a routine analytic technique.

Liquid chromatography–mass spectrometry (LC-MS), gas chromatography–mass spectrometry (GC-MS), and nuclear magnetic resonance (NMR) are powerful tools for molecular structure identification. For example, Maruyama *et al.* (1998) successfully identified the anxiolytic agents in an oriental herbal formula as magnolol and honokiol by utilizing GC-MS and NMR. LC-MS, GC-MS, and NMR are routinely used for drug metabolism studies in drug discovery, but have not yet been used routinely for quality control of plant drugs.

5.2.4   *Separating the analyte from interfering substances*

Sample preparation with suitable solvents is a procedure of separating the analyte from interfering substances. Fujita *et al.* (1973) used liquid phase methods for separation of magnolol and honokiol from *Magnolia* bark. Yamahara *et al.* (1986) reported a solid phase extraction method for separation of magnolol and honokiol. Chromatographic methods have been designed and developed for separation. More detail on separation will be reviewed with the individual methods.

5.3   Performing qualitative and quantitative analyses

5.3.1   *Total extractives*

Plant drugs show characteristic behavior toward particular solvents. Ethanol and water are the most popular solvents for determination of total extractives, and they are also the only solvents allowed for processing of "organic" products in the natural products industry. The Chinese Pharmacopoeia (1995) and the British Herbal Pharmacopoeia (1996) procedures for determining total extractives are well established. In those procedures, 4.0 or 5.0 g of plant coarse powder is extracted with 100 ml of ethanol or water for 24 h at room temperature with frequent shaking during the first 6 h. Then the solution is filtered quickly and 20 ml of the filtrate is evaporated to dryness and dried at 105 °C to constant weight. The total ethanol or water-soluble extractives is calculated and presented as a percentage of the tested sample. The principle of total extractives for quality control is that the differences between the tested drug and the reference standard plant drug should be no more than 10% when tested in identical condition. The Japanese Pharmacopoeia (1996) stipulates that the total extractives of *Magnolia* bark or *Magnolia* bark powder with 70% ethanol should be no less than 12% (w/w).

5.3.2   *UV/vis spectrophotometry*

The UV/vis absorption spectra of various plant constituents can be determined by comparing diluted plant extract against the solvent blank, and its concentration can be calibrated against a standard substance of known dilution. The commonly used solvents are ethanol, methanol, water, hexane, ether, and petroleum ether.

In the Chinese Pharmacopoeia (1995) the quantitative method for magnolol and honokiol uses a combination of TLC and UV/vis spectrophotometer methods. Reflux 1.0 g of *Magnolia* bark powder with 25 ml of ethanol for 1 h, cool down and filter.

Spot 100 µl of filtrate to the silica gel GF_{254} plate side-by-side with 5 µl of reference standard containing 1 mg/ml of magnolol and 1 mg/ml of honokiol. The TLC plate is developed with a mobile phase of benzene–methanol (9:1) mixture for a distance of 8–15 cm. After the plate is air-dried, the band of magnolol or honokiol is collected under UV 254 nm illumination, with reference to the standards. The collected silica is put into separate tubes. The same area of silica is collected from the TLC plate of the blank. Ten milliliters of ethanol is added to each tube and the silices is extracted for 10 min. The tubes are centrifuged and the supernatants are used as test samples. The absorbance is measured at 294 nm, and the concentrations of magnolol and honokiol are calculated on the basis of the calibration curve. The Chinese Pharmacopoeia (1995) stipulates that *Magnolia* bark contain at least 3.0% of total magnolol and honokiol by this method. The TLC separation or purification of the test samples for UV/vis spectrometry is important in this method, since the ethanol extract contains other constituents that interfere with the absorbance at 294 nm.

5.3.3   Fourier transform infrared (FTIR) spectroscopy

Infrared (IR) or FTIR spectroscopy is predominantly applied to qualitative organic analysis, and it has been demonstrated to be a specific, simple, accurate, and fast method for identification of plant drugs. Samples of plant drug powder mixed with potassium bromide, and plant extract in solutions of chloroform, carbon tetrachloride, or mineral oil can be recorded directly by IR/FTIR. Generally, to reduce interference, the plant drug is pretreated with solvents such as 50% ethanol, methanol, acetone, chloroform, or petroleum ether. The appropriate concentration of plant drug is about 1 ml of solvent to 5 mg of plant drug. After complete evaporation of the solvent, the residue is mixed with potassium bromide for FTIR assay.

Figure 5.3 shows the FTIR spectrum of *M. officinalis* bark pretreated with methanol. For plant drug identification, a comparison of the FTIR patterns of the test sample against the reference standard can provide sufficient information, so analysis of the structural groups and fingerprints is usually unnecessary. Yan *et al.* (1994) have published more than 30 FTIR spectra of *Magnolia* bark and its allied plants, which can be used as a reference database. We believe that with computer technology, an FTIR database for *Magnolia* bark, as well as other plant drugs, will be accessible in the near future. This would allow FTIR methods to gain more popularity and to replace TLC for identification of plant drugs. Since it has the highest specificity for qualitative analysis, it requires only a small amount of sample, and its operation is simpler and faster with easier documentation.

5.3.4   Thin-layer chromatography (TLC)

The Chinese Pharmacopoeia (1995) has a TLC method for identification of *Magnolia* bark with magnolol and honokiol as reference standards. The Japanese Pharmacopoeia (1996) gives a TLC method for identification of *Magnolia* bark with alkaloids as marker components, but no reference standard is used. Table 5.2 provides a summary of the TLC conditions and results.

Xie *et al.* (1993) described an additional TLC method in the Chinese Pharmacopoeia TLC ATLAS, which demonstrated honokiol, magnonol, and β-eudesmol simultaneously. The major modifications are as follows. Reflux 1.0 g of *Magnolia* bark with

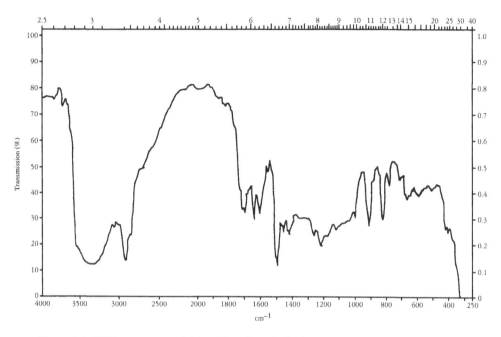

Figure 5.3 FTIR spectrum of *Magnolia officinalis* bark pretreated with methanol.

Table 5.2 Chinese and Japanese Pharmacopoeia TLC methods for identification of *Magnolia* bark

TLC conditions	Chinese Pharmacopoeia 1995	Japanese Pharmacopoeia XIII 1996
Sample preparation	Extract 0.5 g fine powder of *Magnolia* bark with 5.0 ml of methanol for 30 min, filter and spot 5 μl of the filtrate	Extract 0.5 g fine powder of *Magnolia* bark with 10.0 ml of methanol for 10 min, centrifuge, spot 20 μl of the supernatant
Standard preparation	Magnolol and honokiol in methanol at 1.0 mg/ml of each, spot 5 μl each	None
TLC plate	Silica gel GF_{254}	Silica gel GF_{254}
TLC mobile phase	Benzene–methanol 27:1	n-Butanol–water–acetic acid 4:2:1
Distance developed	8 cm	10 cm
Detection	Spray evenly with 1% vanilla-sulfuric acid solution after air-drying of the plate. Heat the plate at 100 °C for 10 min and detect under fluorescent light	Spray evenly with Dragendorff's TS after air drying of the plate and detect under fluorescent light
Major spots and R_f	Compare spots with the references	Yellow spot at R_f 0.3 (alkaloids)

30 ml of ethyl acetate, filter, evaporate the ethyl acetate, dissolve the residue with 0.5 ml of methanol, and use the solution for testing. One percent of NaOH is added to the silica when preparing TLC plates, and the mobile phase is benzene–ethyl acetate (9:1.5). GC and HPLC have recently replaced TLC scanning methods for quantitative analysis of magnolol and honokiol.

5.3.5   Gas chromatography (GC)

GC is the first choice for analyses of volatile oils such as β-eudesmol in *Magnolia* bark. GC methods are also suitable for analysis of small semivolatile molecules, such as magnolol and honokiol.

GC method I

Li *et al.* (1994) reported a GC method for the quantitative analysis of magnolol and honokiol in *Magnolia* bark and *Magnolia* barks processed with traditional methods.

GC CONDITIONS

GC instrument:	HP 5890 II
Column:	OV-101 (2 mm × 2 m)
Column temperature:	220 °C
Carrier gas:	N_2 at 21.3 ml/min
Injection posted temperature:	300 °C
Injection volume:	0.4 μl
Detector:	FID detector (flame ionization)
Retention time:	Magnolol 5 min; honokiol 7 min.

SAMPLE PREPARATION

Accurately weigh 1.0 g of 50-mesh sample powder, put into a small flask, add 5 ml of methanol and extract at room temperature, shake the flask mechanically for 1 h and collect the supernatant into a 50 ml flask, then add 5 ml of methanol to the residue. Repeat the extraction nine more times with a total of 50 ml of methanol. Add methanol to the 50 ml flask to exactly 50 ml, mix well, and use it as the sample solution.

STANDARD CURVE

Magnolol or honokiol is dissolved with ethanol in the following concentrations: magnolol (mg/ml) at 25.69, 21.56, 10.78, 6.289, and 2.516. Honokiol (mg/ml) at 23.40, 17.33, 13.00, 7.800, and 1.950. An aliquot of 0.4 μl of each solution is tested under the GC conditions described above. The calibration curves are:

Magnolol: $Y = 1.710X - 0.8786$, $r = 0.9996$
Honokiol: $Y = 2.196X + 2.595$, $r = 0.9963$

RESULTS

Table 5.3 shows the results obtained from this method.

The authors processed *M. officinalis* bark with the traditional methods and compared their contents of magnolol and honokiol. None of the processing methods, except charring by stir-frying, significantly change the content of magnolol and honokiol.

Table 5.3 Magnolol and honokiol content in processed *Magnolia* bark

Sample	Processing method	Magnolol (%)	Honokiol (%)
Magnolia bark slice	Scrape the outer layer of *Magnolia* bark, wash and soften the bark with water, cut the bark into slices and air dry. The dried slices are tested as control and used for further processing	2.534	1.389
Yellowing	Mix *Magnolia* bark slices well with water (10:1), heat in a pan with cover until the water is absorbed. Stir-fry in the pan with gentle heat until the surface turns to a yellow color	2.187	1.135
Charring	Mix *Magnolia* bark slices with water (10:1), heat in a pan with cover until the water is absorbed. Stir-fry in the pan with mild heat until the surface turns to dark brown	1.378	0.491
Yellowing with ginger juice	The same as yellowing except the water is replaced with fresh ginger juice	2.257	1.147
Yellowing with wine	The same as yellowing except the water is replaced with wine	2.172	1.014
Yellowing with vinegar	The same with yellowing except the water is replaced with vinegar	2.009	0.924

The injector temperature is maintained relatively high to evaporate magnolol and honokiol efficiently. The amounts of magnolol and honokiol used for standard curve preparations are high.

GC method II

Ye *et al.* (1992) reported a programmed temperature control GC (PTGC) method for the quantitative analysis of β-eudesmol, magnolol and honokiol in *M. officinals* with n-C_{20} as internal standard.

GC CONDITIONS

GC instrument:	Shimadazu GC-RIA
Column:	5%SE-30/Chromosorb W (DMCS), 1.6 m × 0.3 mm glass column
Column temperature:	Programmed from 150 °C to 190 °C at 5 °C/min
Injection posted temperature:	250 °C
Carrier gas:	N_2 at 60 ml/min
Detector:	FID detector

REFERENCE STANDARD CURVE

Internal standard: Accurately weigh 22.67 mg of n-C_{20} and dissolve with 10.0 ml of dehydrated ethanol; mix well.

Reference standard: Accurately weigh β-eudesmol 0.92 mg, magnolol 5.43 mg and honokiol 7.96 mg and put into individual 1 ml flasks. Add dehydrated ethanol to exactly 1 ml, dissolve, and mix well.

Calibration curves: Accurately transfer 200 μl of internal standard into a series of small tubes; add different volume of reference standard solution into each tube. Add dehydrated ethanol to make a total volume of 600 μl, mix well. 1.0 μl of the mixture is used per injection.

The calibration curves are:

β-Eudesmol: $Y = 1.305X - 0.016$, $r = 0.9995$
Magnolol: $Y = 1.114X - 0.052$, $r = 0.9996$
Honokiol: $Y = 1.127X - 0.078$, $r = 0.9998$

where, Y represents the peak area of reference standard (A_r) divided by the peak area of internal standard (A_i), and X is the concentration of reference standard. The linear range is 0.096–0.48 mg/ml for β-eudesmol; 0.27–1.36 mg/ml for magnolol; and 0.398–1.99 mg/ml for honokiol. The %RSD values are 0.33%, 0.58%, and 0.73% for β-eudesmol, magnolol, and honokiol, respectively.

The authors also compared sample preparation methods of cool ethanol extraction for 24 h and ultrasonic treatment for 10 min. The ultrasonic method gave comparable results for the three tested constituents, and the authors recommend the ultrasonic method for sample preparation.

RESULTS

The test results obtained using this method are presented in Table 5.4.

Internal standard n-C_{20} is used to help identify β-eudesmol, magnolol and honokiol in the tested samples. The flash point of β-eudesmol is lower than that of magnolol and honokiol; programmed increase of the mobile phase temperature allows this method to separate and quantitate β-eudesmol, magnolol, and honokiol simultaneously.

5.3.6   HPLC methods

Figure 5.4 shows the chemical structure of magnolol (5,5-di-2-propenyl-1,1′-biphenyl-2,2′-diol) and honokiol (3,5′-di-2-propenyl-1,1′-biphenyl-2,4′-diol). They are a pair of isomeric neolignans, and have similar affinities for the stationary phase of C_{18}. Acid is often added to the mobile phase to keep magnolol and honokiol in their neutral states.

HPLC method I (Japanese Pharmacopoeia XIII method)

SAMPLE PREPARATION

Weigh accurately about 0.5 g of pulverized *Magnolia* bark, add 40 ml of diluted methanol (7 in 10), heat under a reflux condenser on a water bath for 20 min, cool, and pass through a filter. Repeat the above procedure with the residue, using 40 ml of diluted methanol (7 in 10). Combine the filtrates, add diluted methanol (7 in 10) to make exactly 100 ml, and use this solution as the sample solution.

Table 5.4 The content of magnolol (M), honokiol (H), and β-eudesmol (B) in *Magnolia* bark and its allied plants by GC

Sample	Origin	Content (%)		RSD (%)
Magnolia officinalis Rehd. et Wils.	No detail	M	4.940	0.92
		H	3.769	1.48
		B	0.208	1.05
M. officinalis var. *biloba* Rehd. et Wils.	No detail	M	1.133	1.32
		H	0.141	1.39
		B	Not detected	–
M. officinalis Rehd. et Wils.	Yi Zhen	M	7.595	0.78
		H	3.258	1.23
		B	0.498	1.89
M. officinalis Rehd. et Wils.	Hubei	M	0.775	1.82
		H	0.544	2.08
		B	0.104	1.40
M. officinalis Rehd. et Wils.	Jiangxi	M	1.045	1.64
		H	0.855	0.67
		B	0.116	1.59
M. officinalis Rehd. et Wils.	Guangxi	M	1.048	1.34
		H	0.413	1.31
		B	0.220	1.87
M. officinalis Rehd. et Wils.	Guangxi	M	6.820	0.45
		H	2.820	1.21
		B	0.674	0.72
M. officinalis Rehd. et Wils.	Hunan	M	7.451	1.60
		H	5.273	2.17
		B	0.280	1.28
M. rostrata W.W. Smith	No detail	M	2.505	0.58
		H	3.004	0.31
		B	1.303	0.47
M. rostrata W.W. Smith	No detail	M	1.648	0.61
		H	4.398	1.16
		B	1.865	0.97
M. biondii Pamp.	No detail	M	Not detected	–
		H	Not detected	–
		B	Not detected	–
M. hookeri Culitt. et W.W. Smith	No detail	M	1.021	4.08
		H	Not detected	–
		B	0.7820	1.39

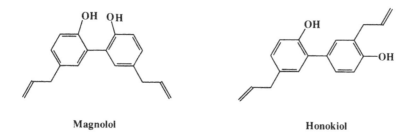

Magnolol   **Honokiol**

Figure 5.4 Chemical structure of magnolol and honokiol.

STANDARD SOLUTION PREPARATION

Dry magnolol for component determination in a desiccator (silica gel) for 1 h or more. Weigh accurately about 0.01 g of it, dissolve in diluted methanol (7 in 10) to make exactly 100 ml, and use this solution as the standard solution.

RESULTS CALCULATION

Perform the test with 10 µl each of the sample solution and the standard solution as directed under liquid chromatography according to the following conditions, and determine the peak areas, AT and AS, of magnolol in each solution. The amount of magnolol in mg = amount of magnolol for component determination × (AT / AS).

HPLC CONDITIONS

Detector:	UV detector at 289 nm
Column:	A stainless-steel column 4–6 mm inside diameter and 15–25 cm long; packed with octadecyl-silanized silica gel (5–10 µm particle size)
Column temperature:	A constant temperature of about 20 °C
Mobile phase:	Water–acetonitrile–glacial acetic acid (50:50:1)
Flow rate:	Adjust the flow rate so that the retention time of magnolol is about 14 min.
Selection of column:	Dissolve 1 mg each of magnolol and honokiol in 10 ml of diluted methanol (7 in 10). Proceed with 10 µl of this solution under the above operating conditions. Use a column giving elution of honokiol and magnolol in that order with resolution between their peaks being not less than 5.
System reproducibility:	When the test is repeated 5 times with the standard solution under the above operating conditions, the relative deviation of the peak area of magnolol is not more than 1.5%.

The Japanese Pharmacopoeia (1996) requires that the *Magnolia* bark contain not less than 0.8% of magnolol using this HPLC method. Harata (1989) reported an identical method for quantitative analysis of magnolol and honokiol from *M. obovata* Thunb. They used a commercial C_{18} column, Nucleosil (4 mm × 25 cm, 5 µm), and reduced the sample volume injected by half. The mean content of magnolol was 2.70% and of honokiol was 0.64% from *M. obovata* Thunb. by their method.

HPLC method II (RP-HPLC with photodiode-array UV detection)

Tsi and Chen (1992) published a HPLC method for qualitative and quantitative analysis of magnolol and honokiol from *M. officinalis*. They provided valuable information on solvent selection for extraction of magnolol and honokiol from *Magnolia* bark as well.

SAMPLE PREPARATION

Reflux 0.5 g of magnolia fine powder with 50 ml of different solvents (water, 99.5% ethanol, 50% ethanol, hexane, 0.1 M HCl in water, and 0.1 M NaOH in water) for 15 min at the boiling point; repeat with the same as different solvent. Combine the filtrate into a 100 ml volumetric flask and add the same solvent to exactly 100 ml. Mix well; 5.0 μl of the sample solution is used per test.

HPLC CONDITIONS

HPLC instrument: Waters 510
Column: Nucleosil C_{18} (4 mm × 25 cm, 7 μm)
Mobile phase: Acetonitrile–water–phosphate (65:35:0.1), pH 2.4–2.7
Flow speed: 1.0 ml/min
Column temperature: Ambient
Detector: Photodiode array UV detector at 209 nm for magnolol and
                        218 nm for honokiol

STANDARD CURVE

Dissolve the appropriate amounts of honokiol or magnolol with methanol, and inject 0.25, 0.5, 1.0, and 2.0 μg for analysis. The calibration curves based on peak area and concentrations are:

Honokiol: $Y = 0.0862X + 0.0028$, $r = 0.999$
Magnolol: $Y = 0.0655X + 0.0057$, $r = 0.999$

Linear relationships for both magnolol and honokiol are obtained at the ranges 5×10^{-3} to 2.0 μg.

RESULTS

Table 5.5 shows the results from the same *Magnolia* bark sample extracted with different solvents.

Hexane and 0.1 M NaOH in water were better solvents for extraction of magnolol and honokiol than water or 0.1 M HCl in water, while the yield with 99.5% ethanol or 50% ethanol was about 15% lower than with hexane. This method uses a photodiode

Table 5.5 Comparison of different solvents for sample preparation of *Magnolia officinalis* bark using magnolol and honokiol as markers by HPLC (*n* = 3, RSD < 3%)

Solvent	Magnolol (mg/g)	Honokiol (mg/g)
Water	2.29 ± 0.89	0.59 ± 0.09
99.5% Ethanol	75.24 ± 3.48	19.13 ± 0.62
50% Ethanol	77.67 ± 0.69	17.27 ± 0.21
Hexane	89.87 ± 3.27	22.11 ± 1.84
0.1 M HCl	1.70 ± 0.73	0.57 ± 0.12
0.1 M NaOH	84.01 ± 1.73	18.29 ± 0.27

array detector (PAD), which can monitor different wavelengths during the chromato-graphic process. A PAD ensures that each peak is monitored at its wavelength of maximum absorbance.

HPLC method III (reversed-phase ion-paired method)

Ueda *et al.* (1993a,b) developed an ion-paired HPLC method for rapid and precise simultaneous determination of magnolol and honokiol in 24 Kampo formulas contain-ing *Magnolia* bark.

HPLC CONDITIONS

Column:	TKSgel ODS-120Å (4.6 mm × 250 mm)
Column temperature:	40 °C
Mobile phase:	Water–acetonitrile (4:6) solution containing 10 mM tetra-n-amylammonium bromide. Adjust pH to 4.0 with phosphate.
Flow speed:	1.0 ml/min
Detector:	UV detector at 290 nm
Sample volume:	10 µl

STANDARD CURVES

Magnolol: $Y = 0.02262X + 7.06222 \times 10^{-5}$, $r = 0.9999$
Honokiol: $Y = 0.02260X - 8.27601 \times 10^{-5}$, $r = 0.9998$

SAMPLE PREPARATION

Boil one day's dosage of plant drugs with 600 ml of water until about 300 ml remains, filter and then cool the solution to room temperature. Add water to the decoction to a total volume of 300 ml. Filter the decoction with a 0.45 µm syringe filter; 10 µl of the filtrate is used for HPLC analysis. Figure 5.5 shows some of the HPLC results.

Ueda *et al.* (1993a,b) used a photodiode array detector and investigated the purity of the magnolol and honokiol peaks. Figure 5.6 shows the three-dimensional chro-matogram spectra of magnolol and honokiol in the reference standard, *Magnolia* bark and "Heii-san", a Kampo formula with six plant drugs including *Magnolia* bark.

Ion-pair chromatography is a technique for analysis of strong acids or strong bases using reversed-phase columns. In this technique, the pH of the mobile phase is adjusted so as to encourage ionization of the sample; for acids pH 7.5 is used, and for bases pH 3.5 is common. In this method, the mobile phase is adjusted to pH 4.0, which keeps magnolol and honokiol at neutral status. The ionization occurs between interfering bases and the ion-pair agent tetra-n-amylammonium bromide (TAA), which changes the retention time and overcomes the poor separation of magnolol and honokiol in the Kampo formula samples. The authors used a single-wavelength UV detector for HPLC analysis and used a photodiode array detector (PAD) for peak iden-tification and for peak homogeneity determination. A single-wavelength detector has higher sensitivity than a variable-wavelength detector or PAD. Variable-wavelength detectors and PAD are advantageous because more than one wavelength can be mon-itored during a single run. A PAD is able to use both retention times and UV spectra

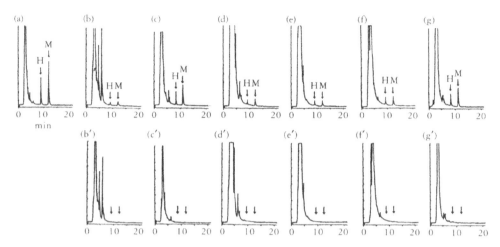

Figure 5.5 Chromatograms of oriental pharmaceutical decoctions. Peaks: H, honokiol; M, magnolol.
(a) *Magnolia* bark; (b) Keisikakoubokukyounin-to; (b′) *Magnolia* bark-deficient keisikakoubokukyounin-to; (c) Koubokushoukyouhangenin-jinkanzou-to; (c′) *Magnolia* bark-deficient koubokushoukyouhangeninjinkanzou-to; (d) Saiboku-to; (d′) *Magnolia* bark-deficient saiboku-to; (e) Shinpi-to; (e′) *Magnolia* bark-deficient shinpi-to; (f) Hangekouboku-to; (f′) *Magnolia* bark-deficient hangekouboku-to; (g) Heii-san; (g′) *Magnolia* bark-deficient heii-san.

to aid in peak identification, and is able to determine the purity of a peak. A PAD scans the whole spectrum within one second, and provides two- or three-dimensional chromatograms. Currently, the sensitivity of variable-wavelength detectors and PADs has been greatly improved, and they are replacing single-wavelength detectors. A PAD is useful for confirming the HPLC separation and identification of the separated peaks.

Yamauchi *et al.* (1996) further developed the ion-paired method to determine five major constituents from four plant drugs simultaneously (hesperidin from Chenpi, 6-gingerol from Ginger, glycyrrhizin from Licorice, and magnolol and honokiol from *Magnolia* bark). The major modification was that a gradient mobile flow was employed for better separation, and a photodiode array detector was used to record chromatograms at multiple wavelengths.

HPLC method IV (reversed-phase ion-paired method for analysis of alkaloids in Magnolia *bark)*

HPLC CONDITIONS

HPLC instrument:	Waters 510
Detector:	UV detector at 282 nm
Column:	Nucleosil C_{18} (300 mm × 4 mm, 5 μm)
Column temperature:	Ambient
Mobile phase:	0.003 mol/l sodium octylsulfonate in water–acetonitrile (74:26) mixture
Flow speed:	1.0 ml/min

(a)

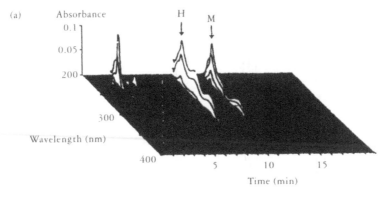

(b)

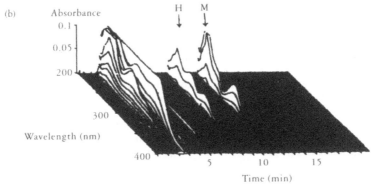

(c)

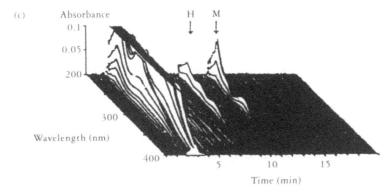

Figure 5.6 Three-dimensional chromatogram spectra. Peaks: H, honokiol; M, magnolol.
(a) Standard honokiol and magnolol;
(b) *Magnolia* bark;
(c) Heii-san.

STANDARD CURVE

Weigh magnocurine, salicifoline, and magnosprengerine 0.5 mg/each; dissolve with 26% acetonitrile in water (pH 3.5) and dilute to 2.0 ml. Inject 0.5, 1.0, 1.5, 2.0, 3.0, and 4.0 μl each for analysis. Draw a standard curve using the concentration as X axis and peak area as Y axis. All three tested components have a linear relationship and the RSD was <1.74%.

SAMPLE PREPARATION

Weigh 0.2 g of fine powder and put into a test tube; add 5 ml of 26% acetonitrile in water (pH 3.5). Extract with ultrasonication for 25 min, centrifuge, and use the supernatant as test sample; 0.5–3.0 μl is used per test.

RESULTS

The tested *M. officinalis* contained ca. 0.06–0.225% of magnocurine; ca. 0.039–0.162% of salicifoline; and ca. 0.018% of magnosprengerine.

5.4 Evaluation (HPLC method validation)

The papers reviewed provide valuable information for qualitative and quantitative analysis of *Magnolia* bark and its allied plants. Instead of comparing the test methods and evaluating their results, we present an HPLC method development for measurement of magnolol and honokiol from *Magnolia* bark and the method validation in detail.

5.4.1 *Determination of magnolol and honokiol in* Magnolia officinalis *and* Magnolia obovata *by HPLC with UV detection*

Introduction

A reversed-phase HPLC method has been developed to simultaneously determine the magnolol and honokiol content in *M. officinalis* and *M. obovata*. The validation of linearity, accuracy, limit of detection, limit of quantitation and precision is described in this study.

Materials

Honokiol and magnolol were purchased from National Institute for the Control of Pharmaceutical and Biological Products, Beijing China. *M. officinalis* Rehd. et Wils bark was purchased at a TCM Pharmacy in Xian, China and identified by TLC. *M. obovata* bark was a gift from Professor Yuji Maruyama, Gunma University, Japan.

Reagents

The methanol, ethanol, and acetonitrile were HPLC grade and were obtained from EM Science (Darmstadt, Germany). The acetic acid, 99.7%, was reagent grade and also obtained from EM Science (Darmstadt, Germany). High-purity water was obtained through a Mili-Q water purification system from Millipore (Bedford, MA, USA).

Instrumentation

Mechanical shaking was performed with a Burrell Scientific (Pittsburgh, PA, USA) Wrist Action Shaker Model 75. The HPLC system used was a Hewlett Packard (Palo Alto, CA, USA) Model 1100 equipped with an autosampler, a photodiode array detector, and a computer running HP Chemstation for data collection and integration.

Standard preparation

A stock solution of the standard was prepared by accurately weighing about 5 mg each of honokiol and magnolol into a 10 ml volumetric flask. Approximately 7 ml of methanol was added to each flask and the solution was sonicated for 15 min. After the solution had cooled to room temperature, it was diluted to volume with methanol and mixed thoroughly. Dilutions from the stock solution that were 2, 5, 25, and 120 times more dilute than the original were created. Calibration curves were constructed for each compound based on three injections of each solution.

Extraction study

To compare the efficiency of various solvents in the extraction of the compounds of interest, samples of each species were prepared with 25%, 50%, 75% ethanol in water (v/v), and 100% ethanol and with 25%, 50%, 75% methanol in water (v/v), and 100% methanol following the method below.

Sample preparation for precision study

Approximately 5 g of the ground bark was put into a 125 ml Erlenmeyer flask. Approximately 170 ml of solvent was added and the flask was shaken mechanically for 24 h. Each solution was vacuum filtered and transferred to a 200 ml volumetric flask. The sample was then diluted to volume with the solvent and mixed thoroughly. Each of the solutions was filtered into an HPLC vial and capped.

Sample preparation for recovery study

A 5 times dilution of the sample preparation described above was spiked with concentrations of the standards equal to 50%, 100% and 150% of the magnolol and honokiol already contained in the sample.

Chromatographic conditions

Column:	Waters Symmetry C_{18} (4.6 × 250 mm, 5 μm)
Column temperature:	30 °C
Mobile phase:	Acetonitrile–water: (60:40) and 0.1% acetic acid (v/v)
Flow rate:	0.80 ml/min
Detection:	UV detector at 290 nm

Results and discussion

Characterization of honokiol and magnolol in Magnolia *bark by HPLC-UV*: Honokiol and magnolol were identified by direct comparison of their retention times with those of authentic samples. Their identities were confirmed by their UV spectrum. Honokiol had a retention time of 10.9 min and magnolol had a retention time of 15.1 min. A typical chromatogram of the standard compared to those of the samples is shown in Figure 5.7. Figures 5.8 and 5.9 show the UV spectrum of honokiol and magnolol respectively.

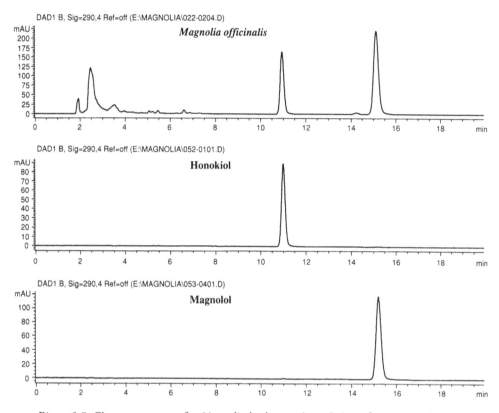

Figure 5.7 Chromatograms of a *Magnolia* bark sample and the reference standards.

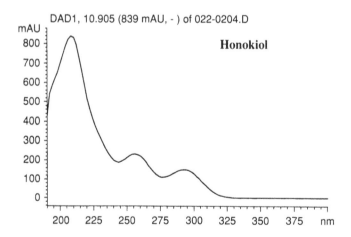

Figure 5.8 UV spectrum of honokiol.

Linearity of calibration curve: The calibration curves for the standards of honokiol and magnolol were constructed from five data points each. Tables 5.6 and 5.7 show the linearity study results and concentrations used. Both compounds showed excellent linearity with correlation coefficients of 1.0000.

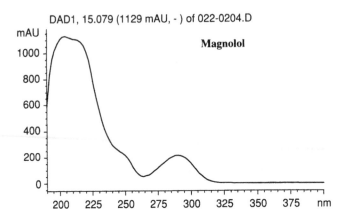

DAD1, 15.079 (1129 mAU, -) of 022-0204.D

Figure 5.9 UV spectrum of magnolol.

Table 5.6 Linearity study results for honokiol

Concentration (µg/ml)	Peak area (MAU·s)[a]	Response factor (µg/ml/MAU·s)[a]
4.07	42	0.09690
20.34	208	0.09779
101.70	1045	0.09732
254.25	2596	0.09794
508.50	5183	0.09811
Average		0.09761
RSD (%)		0.507

Calibration equation
Slope	10.19
Intercept	3.3233
Correlation coefficient	1.0000

[a] MAU, million absorbance units.

Table 5.7 Linearity study results for magnolol

Concentration (µg/ml)	Peak area (MAU·s)[a]	Response factor (µg/ml/MAU·s)[a]
5.06	50	0.1012
25.31	249	0.1016
126.56	1252	0.1011
316.40	3109	0.1018
632.80	6203	0.1020
Average		0.1015
RSD (%)		0.379

Calibration equation
Slope	9.8002
Intercept	4.5338
Correlation coefficient	1.0000

[a] MAU, million absorbance units.

Extraction study: As a result of this extraction study, methanol was found to be the best solvent for extracting both honokiol and magnolol from *M. officinalis* and *M. obovata*. Methanol was subsequently used in all samples prepared to validate this method. A solution of 75% ethanol in water (v/v) was also found to be a good extraction solvent and would be appropriate as the extraction solvent in a production setting. Figures 5.10, 5.11, and 5.12 show the results of the extraction study for honokiol, magnolol, and the total, respectively.

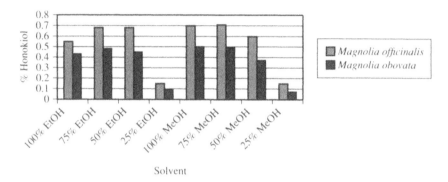

Figure 5.10 Extraction study of honokiol.

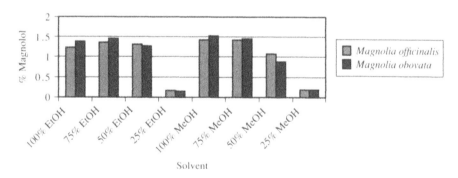

Figure 5.11 Extraction study of magnolol.

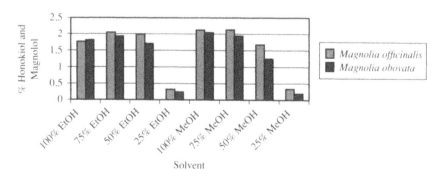

Figure 5.12 Extraction study results for the total content of honokiol and magnolol.

Table 5.8 Precision study results for *Magnolia officinalis*

Injection no.	Honokiol (%)	Magnolol (%)	Total (%)
1	0.72	1.37	2.09
2	0.73	1.38	2.11
3	0.73	1.38	2.12
4	0.74	1.39	2.12
5	0.74	1.39	2.13
6	0.74	1.39	2.13
7	0.74	1.39	2.13
Mean	0.73	1.38	2.12
RSD (%)	1.07	0.57	0.69

Table 5.9 Precision study results for *Magnolia obovata*

Injection no.	Honokiol (%)	Magnolol (%)	Total (%)
1	0.50	1.45	1.95
2	0.50	1.46	1.95
3	0.50	1.46	1.95
4	0.51	1.46	1.96
5	0.50	1.46	1.95
6	0.50	1.45	1.95
7	0.50	1.46	1.96
Mean	0.50	1.46	1.95
RSD (%)	0.00	0.33	0.25

Precision: Seven replicate injections of a *M. officinalis* sample and of a *M. obovata* sample were performed. The data are listed in Tables 5.8 and 5.9 as percent (w/w) of magnolol and honokiol in the ground bark. The relative standard deviation (RSD) for the determination of the sum of magnolol and honokiol is 0.69% for *M. officinalis* and 0.25% for *M. obovata*, indicating excellent sample injection precision and reproducibility of this method.

Six samples of each species of *Magnolia* bark were prepared and analysed. The results are shown in Tables 5.10 and 5.11 as percent (w/w) of honokiol and magnolol in the bark powder. The RSD for the calculation of total content was 2.04% for *M. officinalis* and 1.68% for *M. obovata*, indicating excellent repeatability and precision in the sample preparation method.

Table 5.10 Sample preparation repeatability data for *Magnolia officinalis*

Sample no.	Honokiol (%)	Magnolol (%)	Total (%)
1	0.74	1.39	2.13
2	0.71	1.35	2.06
3	0.73	1.39	2.12
4	0.71	1.34	2.05
5	0.69	1.33	2.02
6	0.71	1.36	2.08
Mean	0.72	1.36	2.08
RSD (%)	2.46	1.86	2.04

Table 5.11 Sample preparation repeatability data for *Magnolia obovata*

Sample no.	Honokiol (%)	Magnolol (%)	Total (%)
1	0.50	1.46	1.96
2	0.50	1.48	1.98
3	0.49	1.44	1.93
4	0.50	1.47	1.97
5	0.49	1.40	1.89
6	0.49	1.45	1.95
Mean	0.50	1.45	1.95
RSD (%)	1.11	1.95	1.68

Table 5.12 Recovery data for *Magnolia officinalis*

Sample	Average recovery honokiol (%)	Average recovery magnolol (%)	Average recovery total (%)
50% Spike	98.11	99.06	98.55
100% Spike	97.26	97.66	97.38
150% Spike	97.62	96.66	96.74
Mean	97.66	97.79	97.56
RSD (%)	0.44	1.23	0.94

Table 5.13 Recovery data for *Magnolia obovata*

Sample	Average recovery honokiol (%)	Average recovery magnolol (%)	Average recovery total (%)
50% Spike	97.40	98.51	98.22
100% Spike	96.15	97.58	97.45
150% Spike	96.44	97.89	97.52
Mean	96.66	97.99	97.73
RSD (%)	0.68	0.48	0.44

Accuracy and recovery: The accuracy and recovery for the bark were evaluated by investigating three different concentrations; approximately 50%, 100%, and 150% of the marker compounds in a typical sample preparation were added to a sample. Spiked samples were prepared in triplicate at each concentration level. The results are listed in Tables 5.12 and 5.13. The average recovery of magnolol and honokiol was 97.56% for *M. officinalis* and 97.73% for *M. obovata*. The results indicate that the accuracy of the method is excellent.

Limits of detection and quantitation: The limit of detection is defined as the concentration at which the signal-to-noise ratio is 3:1. The limit of quantitation is defined as the concentration at which the signal-to-noise ratio is 10:1. To determine these quantities, the most dilute standard solution was successively diluted and the diluted solutions were run with the HPLC method until the signal-to-noise ratios dropped to the designated levels. The limit of quantitation was found to be ca. 0.33 μg/ml for honokiol and ca. 0.41 μg/ml for magnolol; and the limit of detection was found to be ca. 0.13 μg/ml for honokiol and ca. 0.16 μg/ml for magnolol.

The ultrasonic method has been proven to be fast, effective, and precise for *Magnolia* bark sample preparation (Ye *et al.*, 1992; Cei *et al.*, 1988). Alcohol extraction of magnolol and honokiol at room temperature was more effective than at the boiling point (Cei *et al.*, 1988). We used an ultrasonic method for standard preparation and shaking extraction at room temperature for test sample preparation. Our accuracy, recovery, and extraction studies indicate that methanol and 75% ethanol are efficient solvents for sample preparation of *Magnolia* barks, we recommend using methanol for laboratory sample preparation and using 75% ethanol for industrial extraction of *Magnolia* bark.

The λ_{max} for magnolol is is 209 nm, and that for honokiol is 218 nm (Figures 5.8 and 5.9). Tsai and Chen (1992) used a PAD and monitored each peak at its λ_{max}. We set our PAD at a single wavelength of 290 nm, since this wavelength gave good signals for both components and had less interference than lower wavelengths. The wavelength of 289 nm is used by the Japanese Pharmacopoeia (1996). The American Chemical Society guidelines define the limit of detection (LOD) as the concentration at which the signal-to-noise ratio is 3:1, which is the smallest peak that can be judged confidently to be a peak. The smallest detectable peak is too small for accurate quantitation. The limit of quantitation (LOQ), which is the smallest peak whose area can be measured with accuracy, is measured at the concentration at which the signal-to-noise ratio is 10:1. The LOQ was ca. 0.41 µg/ml for magnolol, and ca. 0.33 µg/ml for honokiol in our method. These are ca. 10^5 times lower than the contents of magnolol and honokiol, respectively, in *Magnolia* bark, which indicates that our method is sensitive and there is plenty of room for dilution in sample preparation.

An HPLC assay method for the determination of the honokiol and magnolol content in bark from *M. officinalis* and *M. obovata* has been developed and validated. The assay has been shown to be linear, accurate and precise.

5.5   Discussion

Quality control of *Magnolia* bark and its preparation involves two major concerns: establishment of product specification, and testing the sample to make sure it matches the product specification. Japanese native *M. obovata* was officially accepted as the genuine species in Japan before the Chinese native *M. officinalis* Rehd. et Wils and *M. offlcinals* var. *biloba* Rehd. et Wils were added to the pharmacopoeia in 1996. Generally *M. officinalis* has a higher content of magnolol and honokiol than the Japanese native *M. obovata*. The pharmacopoeia's specifications and its testing methods are changeable, they are updated every five years to reflect the availability of new scientific data. The content of magnolol and honokiol of *M. officinalis*, according to the reviewed articles, varied about 10 times, which indicates the variation in quality of the tested samples. The *M. officinalis* we purchased from a TCM pharmacy in China contained about 2.1% of total magnolol and honokiol, which is below the Chinese Pharmacopoeia specification. Clearly quantitative analysis is essential for QA/QC of *Magnolia* bark, and standardizing *Magnolia* bark and its preparations to a certain amount of magnolol and honokiol assures the product quality.

M. rostrata W.W. Smith contains magnolol and honokiol, and matches the chemical specification of magnolia barks required by the Japanese and Chinese pharmacopoeias. Unlike *M. officinalis* or *M. obovata*, *M. rostrata* has a higher content of honokiol than magnolol. *M. rostrata* has been used as an alternative to genuine species of *Magnolia* barks in China. For anti-anxiety applications, *M. rostrata* might be superior to the genuine species, since Kuribara *et al.* (1999a,b) have identified that honokiol, but not magnolol, is the active compound for the anti-anxiety effects and CNS sedation. *M. sprengeri* Pamp., *M. biondii* Pamp., *M. denudata* Desr., and *M. liliflora* Desr. contain no magnolol or honokiol: they should be classified as adulterants of *Magnolia* bark.

Traditionally *Magnolia* bark was processed with water, ginger juice, wine, and vinegar. Currently the ginger juice yellowing method is recommended by the Chinese Pharmacopoeia (1995). The processing was believed to improve the taste of *Magnolia* bark; to improve its therapeutic effects; and to improve long-term storage by killing contamination from bacteria and insects. The processing methods, except the stir-fry charring, did not change the content of magnolol and honokiol. We speculate that the processing improves the extraction of magnolol and honokiol out of the bark by water, since the TCM decoction uses water as the solvent, and extraction studies (Tsai and Chen, 1992) indicated that water is not an effective solvent for magnolol and honokiol extraction.

Magnolia leaves contain one-fifth of the magnolol and honokiol found in the bark (Zheng *et al.*, 1999). It is not known whether the leaf has the same therapeutic effects as the bark. Leaf extract may match the chemical specification of bark extract in the marker compounds of magnolol and honokiol. To distinguish the plant part(s) used, fingerprint characteristics, such as the ratio between magnolol and honokiol, TLC, FTIR, or other unique markers of the bark should be included in the specification of *Magnolia* bark extract. *Magnolia* leaf is a better source than *Magnolia* bark since it can be harvested annually with almost no harm to the tree. More chemical and pharmacological studies are needed to support its application.

Chromatography is described and measured in terms of four major concepts: capacity, efficiency, selectivity, and resolution. The capacity and selectivity of the column are variables and are controlled largely by the column manufactures. The chromatographer can control the efficiency and resolution to some extent. Weston and Brown (1997) have described the principles and practice of HPLC and CE in detail. Instrument manufactures are also good sources for updated technical information.

This chapter has focused on routine laboratory needs for *Magnolia* bark QA/QC. There are advantages and disadvantages to each method reviewed. Usually a combination of different methods is needed to accomplish the QA/QC of magnolia bark and its preparation. Each laboratory has to select its own test methods and set up its own QA/QC procedures to meet the requirement of its specifications and Good Laboratory Practice. Currently we prefer TLC for qualitative analysis and use HPLC for quantitative analysis of *Magnolia* bark and its preparations. We have provided the HPLC method development and its validation in detail because we hope that the detailed information will help readers to develop their own methods for plant drugs.

Acknowledgments

We thank Mr Bob Green of Nutratech for his full support with preparation of this article, and thank Professor Zhong-wen Li of Hunan TCM College in China and Professor Yuji Maruyama of Gunma University in Japan for their cooperation in collecting reference articles and test samples of *Magnolia* barks.

References

British Herbal Pharmacopoeia (1996) *Method of Analysis*, 4th edn, pp. 195–97. London: British Herbal Medicine Association.

Cei, J.F., Zhang, G.D. and Song, W.Z. (1988) Analysis of *Magnolia* bark and its allied plants I: Analysis of magnolol and honokiol in magnolia bark. *Yao Wu Fen Xi (Drug Analysis)*, 8, 274–276.

Chen, H.Q., Wang, Y.Z., Wang, W.X., Liu, G.F., Li, H.M., Guo, L. and Yuan, Z. (1997) *Quantitative Analysis of Active Components of Common Chinese Herbs*, pp. 464–477. Beijing: People's Health Publisher.

Chinese Pharmacopoeia (1995) *Cortex Magnolia Bark*, pp. 218–220. Guangzhou: Guangdong Science and Technology Publisher.

Fujita, M., Itokawa, H. and Sashida, Y. (1973) Studies on the components of *Magnolia obovata* Thunb. III. Occurrence of magnolol and honokiol in *M. obotata* and other allied plants. *Yakugaku Zasshi*, 93, 429–434.

Japanese Pharmacopoeia (1996) *Magnolia Bark*, 13rd edn., pp. D345–350. Tokyo: Hirugawa Book Store Publisher.

Harata, M. (1989) *Quantitative Analysis of Active Components of Common Plant Drugs*, pp. 145–147. Tokyo: Hirugawa Book Store Publisher.

Kuribara, H., Kishi, E., Hattori, N., Yuzurihara, M. and Maruyama, Y. (1999a) Application of the elevated plus-maze test in mice for evaluation of the content of honokiol in water extracts of magnolia. *Phytother. Res.*, 13, 593–596.

Kuribara, H., Stavinohova, W. and Maruyama, Y. (1999b) Honokiol, a putative anxiolytic agent from magnolia bark, has no diazepam-like side-effects in mice. *J. Pharm. Pharmacol.*, 51, 97–103.

Li, G.Q., Jia, H.S. and Tang, W.W. (1994) Determination of magnolol and honokiol in cortex *Magnolia officinalis* and its processed samples by GC. *Zhong Yao Chai*, 17, 32–33.

Maruyama,Y., Kuribara, H., Morita, M., Yuzurihara, M. and Weintraub, S. (1998) Identification of magnolol and honokiol as anxiolytic agents in extracts of *saiboku-to*, an oriental herbal medicine. *J. Nat. Prod.*, 61, 135–138.

Su, S.W., Xu, C.Q., Sui, C.H., Sun, R.Q., Wang, Y.J., Liu, H.N. and Shen, X.H. (1992) Analysis of the active components of Chinese plant drug "*Huo-Po*" and its allied plants. *J. Shenyang College of Pharmacy*, 9, 185–189.

Tsai, T.H. and Chen, C.F. (1992) Identification and determination of honokiol and magnolol from *Magnolia officinalis* by high-performance liquid chromatography with photodiode-array UV detection. *J. Chromatogr.*, 598, 143–146.

Ueda, J., Asaka, N., Tanaka, I., *et al.* (1993a) A simultaneous determination of honokiol and mangolol in Oriental pharmaceutical decoctions containing magnolia bark by ion-pair high-performance liquid chromatography. II. *Yakugaku Zasshi*, 113, 894–896.

Ueda, J., Momma, N. and Ohsawa, K. (1993b) A simultaneous determination of honokiol and magnolol in oriental pharmaceutical decoctions containing magnolia bark by ion-pair high-performance liquid chromatography. *Yakugaku Zasshi*, 113, 155–158.

Wang, M.Z., Chen, B.H., Araku, K. and Hirayama, S. (1999) *HPLC Analysis of Common Chinese Herbs*, pp. 251–255, Bejing: Science Publisher.

Weston, A. and Brown, P. (eds) (1997) *HPLC and CE: Principles and Practice*. San Diego: Academic Press.

Xie, P.S., Yan, Y.Z., Lin, Q.L., *et al.* (eds) (1993) *Chinese Pharmacopoeia TLC ATLAS of Traditional Chinese Herb Drugs*, pp. 59–61. Guangzhou: Guang Dong Science and Technology Publisher.

Yamahara, J., Miki, S., Matsuda, H. and Fujimura, H. (1986) Screening test for calcium antagonists in natural products. The active principles of *Magnolia obovata*. *Yakugaku Zasshi*, 106, 888–893.

Yamauchi,Y., Ueda, J. and Ohsawa, K. (1996) A simultaneous determination of various main components in oriental pharmaceutical decoction "heii-san" by ion-pair high-performance liquid chromatography. *Yakugaku Zasshi*, 116, 776–782.

Yan, W.M., Guo, Z.F., Yan, Y.N., Tian, H.K. and Dong, X.Q. (1994) *Identification of the Genuine and Sham of Traditional Chinese Medical Drugs*, pp. 347–381. Beijing, People's Health Publisher.

Ye, C.Y., Feng, Y.Q. and Teng, J.C. (1992) Analysis of β-eudesmol, magnolol and honokiol in *Magnolia officinalis* by PTGC method. *Yao Wu Fen Xi (Drug Analysis)*, 12, 159–162.

Zhang, G.D. (1989) A brief review of chemical studies of the medicinal plant houpo. *Chung Kuo Chung Yao Tsa Chih*, 14, 565–568.

Zhang, Z.P., Hu, Z.D. and Yang, G.J. (1997) Separation and determination of magnolol and honokiol in *Magnolia officinalis* bark by capillary zone electrophoresis. *Mikrochim. Acta*, 127, 253–258.

Zheng, H.Z., Dong C.H. and She, J. (1999) *Modern Study of Traditional Chinese Medicine*, pp. 3280–3304: Beijing: Xue Yuan Publisher.

6 Distribution and Commercial Cultivation of *Magnolia*

Suhua Shi, Yang Zhong and William A. Hoch

6.1 Distribution patterns in *Magnolia*
6.2 Commercial cultivation of *Magnolia*

6.1 Distribution patterns in *Magnolia*

6.1.1 *A brief history of taxonomic studies*

The genus *Magnolia* L. ranges widely in eastern and southeast Asia and in the New World in the southeastern United States, the West Indies, Mexico, and Central America to northern South America. Because of its remarkable discontinuous distribution between eastern Asia and eastern North America, *Magnolia* has attracted the attention of biologists. The first published reference to the similarities of species between eastern Asia and eastern North America appears in a Linnaean dissertation published in the mid-eighteenth century (Halenius, 1750; Fernald, 1931; Boufford and Spongberg, 1983). Thunberg (1784) made a brief mention that many Japanese plants occurred in Europe, America, and the East Indies, but particularly in the northern, and adjacent, vast Chinese region. Among the plants he cited was *Magnolia glauca* L., which he believed to be in the two areas. Later, in his travels in the United States, Luigi Castiglioni noted the affinities of many American plants (including *Magnolia*) that he had observed growing with those in Japan. Since then, many authors, such as Pursh (1814), Nuttall (1818), Gray (1840), Miquel (1867), Engler (1879) and Diels (1900), and many recent workers—Hu (1935), Hara (1966), Li (1952), Wood (1971), Raven (1972), Thorne (1972), Boufford (1992), and Sewell *et al.* (1993)—have studied this distribution pattern in detail. *Magnolia*, as a typical disjunct genus between eastern Asia and eastern North America, and the largest genus in the family Magnoliaceae, has a rather clear geographical distribution. However, uncertainties and controversies about the delimitation of the genus have existed for a considerable time because of the overlap in characters between it and other traditional genera of Magnoliaceae (Dandy, 1927, 1964, 1974, 1978; Keng, 1978; Treseder, 1978; Law, 1984, 1997; Nooteboom, 1985, 2000; Chen and Nooteboom, 1993; Figlar, 2000; Shi *et al.*, 2000).

6.1.2 *Classification and distribution patterns in* Magnolia

6.1.2.1 *Classification systems*

Specific delimitation within *Magnolia* has been a subject of persistent debate and disagreement among taxonomists, botanists, horticulturists, and morphologists for at least the last half of the twentieth century. Recently, molecular systematists have joined the debate (Qiu *et al.*, 1995; Azuma *et al.*, 1999; Shi *et al.*, 2000; Ueda *et al.*, 2000). The number of species in the genus differs according to different classification systems of the family Magnoliaceae. Several systems have been proposed for the Magnoliaceae in recent years. In his last treatment, Dandy (1978) recognized 12 genera in the Magnoliaceae, and about 70–80 species. The extremes were shown by the classification of Nooteboom (2000), where only 3 genera and about 150 species in the genus *Magnolia* were recognized, and Liu (= Y.W. Law) (2000) recognized 16 genera in the family Magnoliaceae and about 90 species in *Magnolia* alone. The differences between these two systems are mainly based on the importance assigned to certain morphological features or to some molecular evidence. In Nooteboom's classification of Magnoliaceae, little significance is attached to the position of the inflorescence and the molecular data, and he places all genera except *Pachylarnax* and *Liriodendron* in his large genus *Magnolia*. Liu's classification system uses the position of the inflorescence and many fine morphological characters to recognize a number of small and even monotypic genera, such as his new genus *Woonyoungia* (Law, 1997) plus the medium-sized genus *Michelia*. Frodin and Govaerts (1996) treated the genus *Magnolia* in broad sense to contain approximately 128 species, including those previously placed in *Alcimandra*, *Aromadendron*, *Dugandiodendron*, *Manglietiastrum*, and *Talauma*. Based on observations of the proleptic branching and other morphological characters, Figlar (2000) concluded that there was no reason to maintain the genus *Michelia* separate from *Magnolia*, and then he treated *Michelia* as a subgenus of *Magnolia* along with the subgenus *Yulania*. Recent molecular data do not support the recognition of many small genera in the family (Azuma *et al.*, 1999; Shi *et al.*, 2000; Ueda *et al.*, 2000).

6.1.2.2   *Distribution of subgenera and sections*

The genus *Magnolia* is both temperate and tropical. Here we have mainly adopted 45 taxa listed in the GRIN (Germplasm Resources Information Network) database (based on Dandy's system) (GRIN database, 2001) plus eight endemic species from China (Law, 1996). Of the 53 species, varieties, and hybrids in the genus *Magnolia*, 45 were based on information from GRIN and 8 were from Flora Reipublicae Popularis Sinicae. According to Dandy's (1978) classification system, the genus *Magnolia* is divided into two subgenera: *Magnolia* and *Yulania*. These were further subdivided into 10 sections (not including section *Maingola*), with their general distribution in eastern Asia and the Americas north of the equator. In each of the subgenera, one section contains species from the temperate parts of both North America and eastern Asia.

Subgenus Magnolia

Seven sections with 29 species and varieties and three hybrids are included in subgenus *Magnolia* based on Dandy's classification. Among these sections, two of them are American (section *Magnolia* and section *Theorhodon*), five are Asian (section *Gwillimia*, section *Lirianthe*, section *Oyama*, section *Gynopodium*, and section *Maingola*), and one occurs both in Asia and America (section *Rytidospermum*).

The section *Magnolia* de Candolle contains only a single species, *M. virginiana*, the type of the genus. It ranges along the Atlantic and Gulf coastal plains in the eastern United States from Massachusetts, where it is rare, to North Carolina and Texas.

The section *Theorhodon* Spach has six evergreen species (GRIN database, 2001). All are trees in tropical America, except for *M. grandiflora*, which is a native of the southeastern United States along the Atlantic and Gulf coastal plains. The tropical series range from eastern Mexico (*M. schiediana* and *M. sharpii*), Guatemala (*M. guatemalensis*), Honduras, Costa Rica, and Panama to the mountains of southeastern Venezuela. Another series is West Indian and contains two species, *M. splendens* and *M. portoricensis*. They range from eastern Cuba, Haiti, and the Dominican Republic to western and eastern Puerto Rico.

The section *Gwillimia* de Candolle is a section with about eight evergreen species ranging from southern China through Indochina to the Philippines. Three species are native to China (*M. championii*, *M. delavayi*, and *M. odoratissima*). *M. championii* ranges from Hong Kong to Guangzhou, Hainan, the islands of Guangdong, Taiwan to Guangxi. *M. odoratissima* is restricted to Yunnan. Three species in the section are temperate. *M. delavayi* is also confined to Yunnan. *M. paenetalauma*, however, has a wider area of distribution, ranging from Guizhou, Hainan, and Guangxi in China to northern Vietnam. *M. coco* has the distinction of being the first Asian *Magnolia* to be grown in China in 1786 (Law, 1990). It has a wider distribution than *M. paenetalauma*, ranging from Zhejiang, Taiwan, Guangdong, Guangxi, and Yunnan to northern Vietnam. The other two species are from Hainan and Vietnam (*M. alboisericea*) and from Yunnan, Burma, Thailand, and Laos (*M. henryi*).

There is only one species, *M. pterocarpa*, in the section *Lirianthe* (Spach) Dandy. This species is a tropical tree distributed from Bhutan to India-Assam, Nepal, Burma, and Thailand.

The section *Oyama* Nakai is a group of deciduous shrubs or small trees with four species confined to temperate eastern Asia. In this section, *M. sieboldii* is the easternmost species, and the oldest one in cultivation. It occurs in Korea, Honshu, Kyushu, and Shikoku in Japan, and northeastern China and Anhui, Guangdong, Liaoning, Sichuan, and Yunnan in China. Two endemic Chinese species are in this section: *M. sinensis* occurs only in western Sichuan, and *M. wilsonii* has a wider range, extending from Sichuan to northern Yunnan and Guizhou. *M. globosa* is the most westward-ranging species in this section, extending from eastern Nepal along the eastern Himalaya to Sichuan, Xizang (Tibet), and extreme northwestern Yunnan in China.

The small Asian section *Gynopodium* Dandy includes a few species ranging from southeastern Xizang, northwestern Yunnan (*M. nitida*) and northeastern Upper Burma, Vietnam, to southeastern China and Taiwan.

The section *Rytidospermum* Spach is the only section occurring in both Asia and America in the subgenus. It contains about nine species of deciduous trees. Three series have been recognized in Dandy's classification system, one Asian and the other two American, and each contains two or three species. The Asian series contains *M. hypoleuca*, *M. officinalis* and *M. hypoleuca*, the last one being a native Japanese species and the easternmost of the Asian species of this section. The Chinese species *M. officinalis* ranges from southern Shanxi to southeastern Gansu, southeastern Henan, western Hubei, southwestern Hunan, central and eastern Sichuan to northern Guizhou. Its variety, *M. officinalis* var. *biloba*, appears to occur wild in eastern China (eastern Anhui, western Zhejiang, Mt. Lushan in northern Jiangxi, southern Hunan, southern and central Fujian, northern Guangdong, and northern and northeastern Guangxi). *M. rostrata* grows in northwestern Yunnan and adjacent southeastern Xizang and in northeastern Upper Burma. One of the American series includes three species, *M. tripetala*, *M. fraseri*, and *M. pyramidata*. *M. tripetala* is indigenous to the Appalachian and Ozark Mountains of the eastern United States (Alabama, Arkansas, Florida, Georgia, southern Indiana, Kentucky, Maryland, Mississippi, North Carolina, southern Ohio, eastern Oklahoma, Pennsylvania, South Carolina, Tennessee, Virginia, and West Virginia), and is now introduced and naturalized to New York and Massachusetts. *M. fraseri* has a more restricted distribution in the southern Appalachian Mountains (northern Georgia, eastern Kentucky, western North Carolina, northwestern South Carolina, western Virginia, and West Virginia). *M. pyramidata* grows on the coastal plain of the southeastern United States (Alabama, northwestern Florida, Georgia, Louisiana, southern Mississippi, southern South Carolina, and southeastern Texas). The other American series comprises two species: *M. macrophylla* is native of the southeastern United States in the southern Appalachian and Ozark Mountains (Alabama, northeastern Arkansas [perhaps extirpated], Georgia, Kentucky, Louisiana, Mississippi, North Carolina, southern and central Ohio, Tennessee, and western Virginia). *M. ashei* is closely related to *M. macrophylla* and occurs only on the coastal plain of northwestern Florida. Three hybrids in this section, *M.* × *thompsoniana* (= *M. tripetala* × *M. virginiana*), *M.* × *wiesneri* (= *M. hypoleuca* × *M. toringo*) and *M.* × *vitchii* (= *M. campbellii* × *M. denudata*) have long been cultivated.

Subgenus Yulania

Subgenus *Yulania* comprises three sections with about 19 species (including one variety) and two hybrids, based on Dandy's system. Two of the three sections are temperate Asian. The other one is another disjunct section between Asia and America in the genus *Magnolia*.

There are approximately nine species in the beautiful garden section *Yulania* (Spach) Dandy, which occur in temperate eastern Asia from the eastern Himalaya to eastern China. *Magnolia denudata* is the oldest species in cultivation and native of eastern China, but cultivated in many regions of the country, such as Anhui, Fujian, Guangdong, Guizhou, Hunan, Jiangsu, and Zhejiang, as well as in Japan. *M. sprengeri* replaces *M. denudata* in western Hubei, western Henan, eastern Sichuan, Guizhou, and Yunnan in China. Two of the other westward-ranging species in Asia are *M. dawsoniana* and *M. sargentiana*. Both were found only in western Sichuan and northern Yunnan and are very rare. Another species, *M. campbellii* is the westernmost representative of the section *Yulania* and ranges from Nepal along the eastern Himalaya (Bhutan, India-Assam, Sikkim, and northern Burma) to western Yunnan and Xizang. Treseder (1978) noted the interesting distribution pattern in this area: "As often happens in species with this distribution (Himalayas to west China) the plants from the two extremes of the range show differences, the significance of which is open to different interpretations". The other two native species of China are *M. amoena* and *M. zenii*. They are apparently closely allied to *M. denudata*, but distributed in eastern China. *M. amoena* occurs in Mt. Tianmushan of northern Zhejiang and *M. zenii* ranges from the neighborhood of Nanjing in southern Jiangsu to Henan. Two more native species of central China were recently treated by Law (1996) in Flora Reipublicae Popularis Sinicae, *M. multiflora* M.C. Wang et C.L. Min (1992) and *M. elliptigemmata* C.L. Guo et L.L. Huang (1992). Both of them have very restricted distributions in central China. *M. multiflora* was collected in Shaanxi and Ningxia, and *M. elliptigemmata* has been found in Hubei and Yunnan. A hybrid, *M.* × *soulangeana* (= *M. denudata* × *M.*

liliflora), also belongs to this section. This hybrid has long been cultivated in Hangzhou (Zhejiang), Guangzhou (Guangdong), and Kunming (Yunnan) in China, and has also been widely cultivated in North America. Its early flowering period, however, makes the flowers susceptible to frost damage in the spring.

The section *Buergeria* (Siebold et Zuccarini) Dandy is another temperate Asian section with about six species listed in the GRIN database and in Flora Reipublicae Popularis Sinicae. Three species are in Japan and Korea, one in eastern China, and the other two in central China. *M. kobus* is the most widely distributed of the three Japanese and Korean species. It occurs in southern Japan and on the coast of southern Korea. *M. salicifolia* has a more restricted distribution than *M. kobus*. It was found on Honshu, Kyushu, and Shikoku in Japan. The third Japanese native species is *M. stellata*. It occurs only on the islands of Honshu and Kyushu. *M. biondii* is the northernmost species in China. Its home is in central China, with a range extending from Gansu, Shaanxi, western Henan and western Hubei to eastern Sichuan. The other native Chinese species is *M. cylindrica*, ranging from southern Anhui to northern Jiangxi, northwestern Zhejiang, and northern Fujian. The newly recognized species *M. pilocarpa* Z.Z. Zhao et Z.W. Xie, has a very narrow distribution in Luotian county, Hubei (Zhao and Xie, 1987). *Magnolia* × *loebneri* Kache (= *M. kobus* × *M. stellata*) has been cultivated in many areas.

The section *Tulipastrum* (Spach) Dandy is the second of two sections having a disjunct distribution between Asia and America. There are two deciduous species and one variety in the section. One Asian species and one American species and its variety have long been in cultivation. The Asian species *M. liliflora* has been cultivated for both medicinal and ornamental uses in China and Japan. It is believed to have originated in eastern China and probably in the temperate region south of the Changjiang (Yangtze) River. The American species *M. acuminata* and its variety *M. acuminata* var. *subcordata* are the most widely distributed *Magnolia* in the Americas. *M. acuminata* is the only species to reach Canada (extreme southeastern Ontario). Geographically, it ranges from the Appalachian and Ozark Mountains of the eastern United States to southern New York.

6.1.3   *The formation of distribution patterns*

6.1.3.1   *Paleobotany of* Magnolia

The emergence of fossils that are convincingly *Magnolia* was in the Early Cretaceous *Pseudofrenelopsis*-Angiosperms Assemblage in what is regarded as Aptian-Albian in age (Tao and Zhang, 1992; Doyle and Hickey, 1976). The earliest fossils of this genus, found in northeastern China (Jilin), have been named *Archimagnolia rostrato-stylosa* Tao et Zhang. The Late Cretaceous was an important stage of angiosperm development. Although reliable fossils of *Magnolia* were found as early as the Early Cretaceous, they were in only a subordinate status in the vegetation because gymnosperms still predominated in the Early Cretaceous floras. Since the Late Cretaceous, however, especially in the Turonian, angiosperms (including magnolias) gradually became dominant (Li, 1995). Many more fossils of magnolias, such as *Magnolia* sp. and *Magnolioxylon* were found in Japan, in the far eastern areas of Russia, and in North American Late Cretaceous floras (Guo, 1986; Bell, 1957; Dilcher and Crane, 1984; Page, 1970). In the Early Tertiary, many fossils of *Magnolia* and *Magnoliastrum*, were found in widespread regions of the world. The fossil sites ranged from Asia (southeastern Asia, Yunnan, Xizang, Shandong and northeastern China) (Bande and Prakash, 1986; Guo and Sun, 1989; Tao and Du, 1982; Tao and Zhang, 1992; Hu and Chaney, 1940), Europe (Tiffney, 1977; Takhtajan, 1974) to North America (Taylor, 1990; Dilcher and Crane, 1984).

Early Cretaceous (Barremian to Albian) *Magnolia* pollen, so-called *Magnolia-Magnolipollis*, was reported from the Early Cretaceous Xinmingbao Group in several locations of Gansu by Hsu *et al.* (1974) and Ye and Zhang (1990), which is earlier than the fossil *Archimagnolia*. Considerable deposits of *Magnolia* pollen have been found in widely scattered strata of Cretaceous and Tertiary age in mainland China (Gansu, Xinjiang, Qinghai, Jilin, Heilongjiang, Yunnan, Guangdong, Sichuan, Hubei, Shaanxi, and eastern China) (Song, 1986; Song *et al.*, 1981, 1992; Tao and Kong, 1973; Li *et al.*, 1978; Sun, 1979; Jiang *et al.*, 1988).

On the basis of current information from megafossils and pollens, the early evolution of the genus *Magnolia* could have taken place during the Early Cretaceous or even earlier. The development of the genus may have reached a rather high level at the time.

6.1.3.2   *Present distribution centers of* Magnolia

The geographical distribution of the species of *Magnolia* is summarized in Table 6.1. This genus is widely dispersed throughout the temperate Northern Hemisphere (Law, 1999). Its greatest concentration of species is in southeastern Asia, in the region extending from the eastern Himalaya eastward to China and southward to Java (Dandy, 1964). A detailed analysis of the geographical distribution of *Magnolia* by areas follows (see Table 6.2).

Asia-Temperate: This area, where the greatest concentration of species occurs and which has been a major center of speciation, is undoubtedly the present distribution center of *Magnolia*. The region is also important as an area of local endemism. Species diversity is greatest from the eastern Himalaya to eastern Asia. There are about 31 species and varieties distributed either widely or locally in this area. These species represent 58.5% of all species in the genus and discussed in this chapter. Three sections of *Magnolia* are restricted to this area: section *Oyama* (subgenus *Magnolia*), section *Yulania* and section *Buergeria* (subgenus *Yulania*).

Asia-Tropical: The diversity of the genus *Magnolia* is not as rich in tropical Asia as in temperate Asia. In total, about eight species and varieties, i.e., 15.1% of the taxa in the genus, are distributed in the Asian tropics. All species of this area belong to subgenus *Magnolia* and almost all of them are in three sections: section *Gwillimia*, section *Lirianthe* and section *Gynopodium*. One species, *Magnolia rostrata*, is in the disjunct section *Rytidospermum*. The section *Lirianthe* is a small, typically tropical Asian section with only one species, *M. pterocarpa*, which ranges within a confined area in Bhutan, India-Assam, Nepal, Burma, and Thailand.

North America: This area has the same species richness for *Magnolia* as the tropical Asian area. About nine species and varieties (17% of all the taxa of *Magnolia*) are divided among four sections and two subgenera. Of these, the distribution patterns of

Table 6.1 Geographical distributions of *Magnolia*

Taxon	Asia			America	
	Temperate Asia	Tropical Asia	China	North America	South America
M. acuminata				+	
M. acuminata var. subcordata				+	
M. albosericea		+			
M. amoena	+		+		
M. ashei				+	
M. biondii	+		+		
M. campbellii	+	+			
M. championii	+		+		
M. coco	+				
M. cylindrica	+		+		
M. dawsoniana	+		+		
M. delavayi	+		+		
M. denudata	+		+		
M. elliptigemmata	+		+		
M. excelsa	+	+			
M. fraseri				+	
M. globosa	+	+			
M. grandiflora				+	
M. guatemalensis					+
M. henryi	+	+			
M. hypoleuca	+				
M. kobus	+				
M. liliiflora	+		+		
M. macrophylla				+	
M. multiflora	+		+		
M. nitida	+	+			
M. odoratissima	+		+		
M. officinalis	+		+		
M. officinalis var. biloba	+		+		
M. paenetalauma	+				
M. pilocarpa	+		+		
M. portoricensis					+
M. pterocarpa		+			
M. pyramidata				+	
M. rostrata	+	+			
M. salicifolia	+				
M. sargentiana	+		+		
M. schiedeana				+	
M. sharpii				+	
M. sieboldii	+				
M. sinensis	+		+		
M. splendens					+
M. sprengeri	+		+		
M. stellata	+				
M. tripetala				+	
M. virginiana				+	
M. wilsonii	+		+		
M. zenii	+		+		
M. × loebneri					+
M. × soulangeana					+
M. × thompsoniana					+
M. × veitchii					+
M. × wiesneri					+

Table 6.2 Geographical distributions and species numbers of subgenera and sections of *Magnolia*

Subgenus/Section	Asia			America	
	Temperate Asia	Tropical Asia	China	North America	South America
Subgenus *Magnolia*	15	6	7	8	4
Section *Gwillimia*	6	2	2		
Section *Lirianthe*		1			
Section *Rytidospermum*	4	1	2	5	
Section *Magnolia*				1	
Section *Oyama*	4	1	2		
Section *Theorhodon*				2	4
Section *Gynopodium*	1	1			
Subgenus *Yulania*	16	1	12	2	
Section *Yulania*	9	1	8		
Section *Buergeria*	6		3		
Section *Tulipastrum*	1		1	2	
Total	31	7	19	9	5

two sections (section *Rytidospermum* and section *Tulipastrum*) are disjunction on different continents in the Northern Hemisphere; one (section *Theorhodon*) ranges widely from South to North America; the fourth section, section *Magnolia*, is confined to North America. This area represents another center of diversity for the genus *Magnolia*.

South America: This area has the smallest number of naturally occurring species in the genus (only five species or 9.4% of the genus). The species of South America are confined to section *Theorhodon* in subgenus *Magnolia*.

China: Although China is probably not a natural geographical distribution area, without doubt it is very important as an area of local endemism, with 19 endemic taxa of *Magnolia* (35.9% of the genus). The endemic species in China are separated into six sections and two subgenera: three sections (section *Gwillimia*, section *Oyama*, and section *Rytidospermum*) in the subgenus *Magnolia* and three sections (section *Yulania*, section *Buergeria*, and section *Tulipastrum*) in the subgenus *Yulania*. They are also separated into two sections with disjunction ranges between eastern Asian and North American. China is therefore the center of the present day distribution of *Magnolia* in Asia.

6.2   Commercial cultivation of *Magnolia*

6.2.1   *Main uses and habits of* Magnolia

Most *Magnolia* species have economic importance (Callaway, 1994; Wiersema and Leon, 1999). They have been extensively introduced and cultivated as ornamental plants (e.g., *M. grandiflora* and *M. coco*), timbers with relatively high quality (e.g., *M. officinalis* var. *biloba*), medicinal plants (e.g., a famous Chinese traditional medicinal material Flos Magnoliae is from *M. liliflora*), and natural resources of stacte, a sweet spice used in making incense, and flavour (e.g., *M. cylindrica*), for a long time (Table 6.3).

Almost all *Magnolia* species are valuable planted ornamentals. For example, *M. grandiflora* is native of the middle and southern sections of Georgia, South Carolina, Alabama, Louisiana, and the upper districts of Florida, USA, and has been extensively planted worldwide. It is a noble urban landscape tree because it is resistant to acid deposition (Adams, 1972; Gumeringer, 1989).

In China, some *Magnolia* species, e.g., *M. elliptigemmata* and *M. pilocarpa*, have become new resources of medicinal Biond Magnolia Flower (*Flos Magnoliae*), besides *M. liliflora* (Guo and Huang, 1992; Zhao and Xie, 1987). Phytochemical studies of *Magnolia* have reported that the leaves, fruits, barks, and woods of many species, e.g., *M. grandiflora*, yield a variety of extracts with potential applications as pharmaceuticals (Adams, 1972; Gumeringer, 1989).

Nearly two-thirds of commercial *Magnolia* species can be used in making furniture products (Priester, 1990). The sapwood of *Magnolia* is creamy white, while the heartwood is light to dark brown, often with greenish to purple-black streaks or patches. The high-quality wood of *Magnolia* is even-textured and moderately heavy, fairly hard and straight grained. It is resistant to heavy shrinkage, is highly shock absorbant, and has a relatively low bending and compression strength. It takes glue well, has a good nailing quality, and stains and varnishes easily (Godfrey, 1988). *Magnolia* wood is also used by the food industry for making cherry boxes, flats, and baskets, and is used for popsicle sticks, tongue depressers, broomhandles, veneers, and venetian blinds, in North America (Duncan and Duncan, 1988; Godfrey 1988).

Magnolia may be important to livestock and wildlife. For example, *M. virginiana* is one of the major forages for deer and cattle (Priester, 1990). Winter use by cattle can account for as much as 25% of their diet (Lay, 1957). White-tailed deer browse the leaves and twigs year-round (Little *et al.*, 1958). The seeds of *M. virginiana* are eaten by gray squirrels and to a lesser extent by white-footed mice, wild turkey, quail, and song birds (Simpson, 1988), and *M. grandiflora* by squirrels, opossum, quail, and the wild turkey (Adams, 1972; Olson *et al.*, 1974).

In addition to economic significance mentioned above, some rare and endangered *Magnolia* species are important in conservation biology. For example, *M. sharpii* is endemic to the state of Chiapas, Mexico, and is known to occur in only five small fragmented populations. *M. schiedeana* is also endemic to Mexico and is listed as threatened with extinction.

The main habits or botanical characteristics of 53 taxa of *Magnolia* are also given in Table 6.3. In general, they are evergreen or deciduous arbors or shrubs. Most arbor species (e.g., *M. denudata* and *M. cylindrica*) are middle-tall trees in evergreen hardwood forests or evergreen deciduous-hardwood mingled forests. Some trees can reach heights of 24 m, with a diameter of 1 m. A wild tree of *M. denudata* found in Mt. Yuntai, Jiangsu, China, is approximately 1000 years old (Law, 1990).

Most *Magnolia* species are temperature-sensitive trees. The North American natives bloom in early summer after the leaves expand, and many of the Asian species bloom in early spring on naked twigs (Elias, 1980). The blooming date of a *Magnolia* species varies according to the climate in which it is located. For the same species in different locations or different species in the same location, the dates may differ by up to 4 or 5 months. The annual breeding periods of *Magnolia* species may also be different. Some species, such as *M. denudata* cultivated in Shanghai and southeastern China, have

Table 6.3 Uses and main habits of *Magnolia*

Taxon	Use[a]	Habit		
		Type	Flower color	Mature height (m)
M. acuminata	O	Deciduous tree	White, yellow	15–24
M. acuminata var. subcordata	O	Deciduous tree	Yellow	10
M. albosericea	O	Evergreen tree	White	3–5
M. amoena	O/T	Deciduous tree	Pink	5–8
M. ashei	O	Deciduous tree or shrub	White	4.5
M. biondii	M	Deciduous tree	White, dark rose	5
M. campbellii	O/M	Deciduous tree	Red, pink	30
M. championii	O	Evergreen shrub or tree	White	1–3
M. coco	O/F	Evergreen shrub	White, yellow	1
M. cylindrica	O/F	Deciduous tree	White	3–5
M. dawsoniana	O/R	Deciduous tree	Rose	10
M. delavayi	O	Evergreen tree	White	6–9
M. denudata	O/M	Deciduous tree	White	15–22
M. elliptigemmata	M	Deciduous tree	White	7–12
M. excelsa	O	Evergreen tree	White	5
M. fraseri	O	Deciduous tree	White	15
M. globosa	O	Deciduous tree	White	3–5
M. grandiflora	O/T/M	Evergreen tree	White	18–27
M. guatemalensis	O/R	Evergreen tree	White	10–20
M. henryi	O	Evergreen tree	White	6.2
M. hypoleuca	O	Deciduous tree	White	5
M. kobus	O/T	Deciduous tree	White	4–6
M. liliiflora	O	Deciduous shrub	Red	1
M. macrophylla	O	Deciduous tree	White	12
M. multiflora	O	Deciduous tree	White	14
M. nitida	O/R	Deciduous tree	White	10
M. odoratissima	O/F	Evergreen tree	White	5–8
M. officinalis	O/M	Deciduous tree	White	10
M. officinalis var. biloba	O/M/R	Deciduous tree	White	8–10
M. paenetalauma	O	Evergreen shrub or tree	White	3–10
M. pilocarpa	M	Deciduous tree	White	12–15
M. portoricensis	O/R	Deciduous tree	White	6
M. pterocarpa	O	Deciduous tree	White	5
M. pyramidata	O	Deciduous tree	Rose	13–20
M. rostrata	O	Deciduous tree	White	25
M. salicifolia	O	Deciduous tree	White	7.5
M. sargentiana	O/R	Deciduous tree	Light red	8–20
M. schiedeana	O/R	Evergreen tree	White	15
M. sharpii	O/R	Evergreen tree	White	10–20
M. sieboldii	O	Deciduous tree	White	3–6
M. sinensis	O	Deciduous tree	Pink, white	3
M. splendens	O/R	Deciduous tree	White	5
M. sprengeri	O/M	Deciduous tree	White	10
M. stellata	O	Deciduous tree	White	5.5
M. tripetala	O	Deciduous tree	White	9
M. virginiana	O/T/F	Deciduous to semi-evergreen tree	White	6
M. wilsonii	O	Deciduous tree	White	3
M. zenii	O	Deciduous tree	White	2–3
M. × loebneri	O	Deciduous tree	White	5
M. × soulangeana	O	Deciduous tree	Pink, white, dark rose	7.5
M. × thompsoniana	O	Deciduous tree	White	6
M. × veitchii	O	Deciduous tree	White	6
M. × wiesneri	O	Deciduous shrub	White	2–4

[a] O = ornamental; M = medicinal; T = timber; F = flavoring/stacte; R = rare.

a relatively long breeding period within a year: flowering in early March, leafing after flower dropping, floral initiation in the end of June, fruit setting in September, leaf color changing in October, and leaf dropping in November. Seed-picking time of *Magnolia* ranges from September to November.

The flowers of *Magnolia* species are terminal, solitary, large, and often showy (Staff of the L.H. Bailey Hortorium, 1976). Pollination is facilitated by beetles feeding in the flowers (Callaway, 1994).

6.2.2 *Environmental factors for growing* Magnolia

6.2.2.1 *Siting*

Many *Magnolia* species with large-sized leaves seem to be wind-tolerant, but their branches are somewhat brittle. Trees growing in sheltered situations are more prone to damage by freak wind gusts than those growing in open situations, where adaptation to environment brings about the development of shorter and stouter branches and sometimes smaller leaves than usual.

In a part-shaded situation, however, colored flowers of *Magnolia* are darker and retain their color better. For example, *M.* × *loebneri* "Leonard Messel" is at its best when it flowers during a sunless period or in a shaded situation. When grown in the open it loses much of its color when warm sunny weather coincides with its flowering season (Treseder, 1978). *M. grandiflora* is also moderately tolerant of shade. It can endure considerable shade in early life but needs more light as it becomes older (Glitzenstein *et al.*, 1986). It will invade pine or hardwood stands and is able to reproduce under a closed canopy. It will not reproduce under its own shade. Once established, it can maintain or increase its presence in stands by sprouts and seedlings that grow up through openings that occur sporadically in the canopy (Adams, 1972). *M. grandiflora* has been migrating onto mesic upland sites and establishing itself, along with associated hardwoods, as part of the climax forest (Myers and White, 1987; Olson *et al.*, 1974).

6.2.2.2 *Soil*

Magnolia likes sour, muck-rich, and well-drained soils, but does not like stagnant water or too high a water table, i.e., waterlogged conditions. Most *Magnolia* species are very tolerant of heavy clay or chalky soils. However, an excess of lime in the soil can cause chlorosis of *Magnolia* leaves and accelerate the depletion of soil humus, because lime aids bacterial activity and breaks down the complex carbon compounds in humus. Similarly, very low potash levels in the soil can induce chlorosis in *Magnolia*. Conversely, an excess of potash may result in salinity and also aggravate magnesium deficiency (Treseder, 1978).

In the United States, for example, *M. grandiflora* grows best on rich, loamy, moist soils along streams and near swamps in the Coastal Plain (Maisenhelder, 1970). It is found on a number of different soils including those in the orders Spodosols, Alfisols, Vertisols, and Ultisols.

6.2.2.3 *Climate*

Most *Magnolia* species are thermophilous and cold-tolerant plants. Lape (1966) reported that *M. acuminata*, *M. kobus*, and *M. stellate* as well as the Loebner hybrids

(*M. kobus* × *M. stellate*) were completely hardy in the winter lows down to −28.9 °C in New York. Leach (1973) also reported a comparison of *Magnolia* species in winter hardiness, at Brookville, Pennsylvania, in the foothills of the Allegheny Mountains in early 1963 (the thermometer went down to −37.2 °C): among 30 species, cultivars, and hybrids investigated, eight were "no injury", twelve "slight injury", two "moderate injury", six "severe injury", and two "killed".

It is known that *M. grandiflora* grows in warm temperate to semitropical climates in the United States (Bennett, 1965). The frost-free period is at least 210 days and is more than 240 days for much of the range. Average January temperatures along the coast are 9–12 °C in South Carolina and Georgia and 11–21 °C in Florida. Coastal temperatures average 27 °C during July. Temperatures below −9 °C or above 38 °C are rare within the species natural range. Annual rainfall averages 1020–1270 mm in the northeastern portion of the range and 1270–1520 mm in other areas (Tripp and Raulston, 1995).

6.2.2.4 Planting

Magnolia may be planted from November to May. Container-grown plants can be planted out during the summer months if reasonable care is taken to saturate thoroughly and avoid any disturbance of their root balls, and to keep them adequately irrigated during dry weather until fully established. In order to retain soil moisture and feed the surface roots of *Magnolia*, a surface-mulch of similar organic materials may also be added after planting.

Deep planting should be avoided for most species because they are surface rooting (Treseder, 1978). However, some species, e.g., *M. grandiflora*, are deep-rooted trees, except on sites with a high water table. Seedlings quickly develop one major taproot. As trees grow, the root structure changes. Trees of sapling stage and beyond have a rather extensive heart root system. Older trees may develop a fluted base with the ridges corresponding to the attachment of major lateral roots.

6.2.2.5 *Factor comparisons between wild and cultivated* Magnolia: *an example*

More than 120 Magnoliaceae species, varieties, and hybrids have been introduced from southern and eastern China, Japan, Vietnam, and the United States, and cultivated in the Magnolias Garden of South China Botanical Garden, Academia Sinica, by Professor Y.-H. Liu (Y.W. Law) and his colleagues for about 30 years. South China Botanical Garden is located in northeastern Guangzhou, Guandong Province (23°10′ N, 113°21′ E) and has an area of 15 ha. Differences in main environmental factors between the original places and the Magnolias Garden have produced different performances of these cultivated plants of the family Magnoliaceae (Liu *et al.*, 1997).

Tables 6.4 and 6.5 show the differences of some climate and soil factors between 10 original places and South China Botanical Garden (SCBG), and the performances of 32 cultivated taxa of the genus *Magnolia* in the garden (Liu *et al.*, 1997). In general, although the environmental conditions between SCBG and the original places are different, the cultivation of *Magnolia* in SCBG has been relatively successful (for example, 62.5% of cultivated species and varieties as well as hybrids have flowered).

Table 6.4 Climate and soil factors in South China Botanical Garden (SCBG) and original places of *Magnolia*

Location	Longitude and latitude	Altitude (m)	Mean annual temperature (°C)	Extreme lowest temperature (°C)	Extreme highest temperature (°C)	Mean annual precipitation (mm)	Relative moisture (%)	Type of rock and soil	pH
SCBG	23°10′N, 113°21′E	30	22.0	0.4	38.0	1610	81	Granite, lateritic red soil	5
Wenshan, Yunnan	23°23′N, 104°13′E	2000	15.0	−7.8	31.0	1400	85	Granite, mountain yellow earth	4.2
Maguan, Yunnan	23°02′N, 104°23′E	1333	16.9	−3.2	31.9	1291	84	Limestone, black earth	7.3
Fangcheng, Guangxi	21°57′N, 108°36′E	50	22.0	1.2	37.2	2103	81	Granite, latosol red soil	5.1
Rongshui, Guangxi	25°05′N, 109°12′E	123	19.3	−2.7	38.6	1892	80	Granite, mountain yellow earth	5.2
Longsheng, Guangxi	25°50′N, 110°00′E	960	14.8	−6.2	32.7	2634	83	Arenaceous, mountain yellow earth	4.9
Ruyuan, Guangdong	24°48′N, 113°15′E	508	16.7	−3.6	36.9	1825	84	Granite, mountain yellow earth	4.7
Ledong, Hainan	18°65′N, 108°70′E	920	26.0	4.0	41.0	1800	80	Granite, red soil	6.7
Linan, Zhejiang	30°00′N, 119°25′E	1507	8.9	−20.6	30.6	1659	85	Limestone, mountain yellow earth	4.9
Hengshan, Hunan	27°11′N, 112°43′E	400	16.2	−4.5	36.0	1900	80	Granite, mountain yellow earth	4.3
Xinning, Hunan	26°15′N, 111°19′E	500	17.0	−6.7	38.8	1330	80	Arenaceous, mountain yellow earth	4.5

Table 6.5 Cultivated *Magnolia* in the South China Botanical Garden, Academia Sinica (data from Liu *et al.*, 1997). The classification of *Magnolia* follows Law (1996)

Taxon	Year of introduction	Original place	Performance
M. acuminata var. *subcordata*	1982	Southeastern USA	Good, flowered
M. albosericea	1974	Hainan, China	Good, flowered
M. amoena	1981	Zhejiang, China	Fair
M. biondii	1981	Hunan, China	Fair, flowered
M. championii	1983	Guangdong, China	Good
M. coco	1981	Guangxi, China	Good, flowered
M. cylindrica	1981	Anhui, China	Poor
M. delavayi	1982	Yunnan, China	Poor, flowered
M. denudata	1981	Zhejiang, China	Fair, flowered
M. denadata var. *pyriformis*	1993	Shaanxi, China	Good
M. globosa	1981	Hunan, China	Fair
M. grandiflora	1979	Southeastern USA	Good, flowered
M. grandiflora var. *lanceolata*	1982	Southeastern USA	Good, flowered
M. hainanensis	1986	Hainan, China	Good
M. henryi	1983	Yunnan, China	Good, flowered
M. kobus	1981	Tokyo, Japan	Fair
M. liliiflora	1982	Zhejiang, China	Fair, flowered
M. liliiflora var. *multiplex*	1986	Fujian, China	Good, flowered
M. obovata	1981	Tokyo, Japan	Fair
M. odoratissima	1983	Yunnan, China	Good, flowered
M. oerissima	1991	Yunnan, China	Good
M. officinalis	1981	Guangxi, China	Fair, flowered
M. officinalis subsp. *biloba*	1977	Jiangxi, China	Fair, flowered
M. paenetatauma	1958	Hainan, China	Good, flowered
M. purpurella	1981	Hunan, China	Good, flowered
M. shangsiensis	1981	Guangxi, China	Good, flowered
M. seiboldii	1984	Liaoning, China	Poor, flowered
M. soulangeana	1977	Zhejiang, China	Poor, flowered
M. sprongeri	1981	Hunan, China	Poor, flowered
M. stellata	1988	Tokyo, Japan	Poor
M. tripetala	1982	Southeastern USA	Poor
M. zenii	1982	Jiangsu, China	Poor

6.2.3 *Management of cultivated* Magnolia

6.2.3.1 *Fertilizer*

Most woody plants have comparatively low nutrient requirements, and cultivated plants of *Magnolia* have only normal requirements for fertilizers. However, phosphate fertilizers should be used in sour soils, especially before the periods of blooming and fruit setting of *Magnolia* (Law, 1990).

When applying fertilizer, it should be borne in mind that the majority of the feeding roots of *Magnolia* are beneath the drip-line of the outer branches and not close to the base of the bush or tree. *Magnolia* species are surface rooting, so on no account should the ground be dug over beneath their branches (Treseder, 1978).

6.2.3.2   Irrigation

Winter drought can cause extensive die-back and mortality of most *Magnolia* species, such as *M. grandiflora*. Irrigation and water retention are also necessary for *Magnolia* during summer. Where watering becomes advisable, the irrigation should be prolonged and copious.

6.2.3.3   Pruning

Magnolia requires no routine pruning but it may be pruned; in particular, newly planted trees may be cut down to ground level. Removal of at least half of the mature leaves is a pretransplanting recommendation for rooted layers of the trees, such as *M. grandiflora* and *M. delavayi*.

Where a strong basal shoot arises on young trees of *Magnolia* growing on their own roots, the original growth may be cut away in March or April. Some species tend to produce long, gangling branches and it may become desirable to shorten these, especially where space is limited. The best times to do this are immediately after flowering, or in July or August (Treseder, 1978).

A simple method of root pruning for some *Magnolia* was suggested by Cyril H. Isaac (Treseder, 1978). In this it is recommended that a circle be marked out about 1 m from the trunk. A sharp spade is then forced down to its full depth to sever all roots in its path. Then, missing a spade's width, the operation is repeated right around the tree. This method may be useful for restricting the growth of the larger types and reducing the time taken for a tree to begin flowering.

6.2.3.4   Diseases and pest control

Most *Magnolia* species have relatively strong disease resistance. Die-back of shoots and branches can be attributed generally to a natural overcrowding of the branches, combined with the relatively large size of most *Magnolia* leaves. Secondary fungal organisms soon prey upon these overcrowded growths and may resemble a primary infection. For example, a number of fomes and polyporus fungi cause heartrot in *M. grandiflora*. Leaf spot, sometimes spreading over the surface of isolated leaves in more-or-less concentric rings, has been attributed to the common gray mold fungus (*Botrytis cinerea*), and also to *Phyllosticta* species. Cankers, occasionally met with in *Magnolia*, are generally attributed to infection by *Phomopsis* species. They rarely occur where the basic requirements of fertile, not-too-alkaline soil, adequate drainage, and suitable climatic conditions prevail (Treseder, 1978).

Moles and occasionally field-mice and ants have been known to jeopardize recently planted *Magnolia* by burrowing beneath their roots and thereby loosening and drying out the soil. It is a good idea to water the compacted area copiously prior to raking over with loose soil or mulching materials.

Mealy-bug will attack *Magnolia* grown perpetually under glass. A simple method of control is to dab paraffin or methylated spirits on to the mealy covering beneath which the insects are hidden. Greenhouse red spider (*Tetranychus telarius*) can cause considerable damage to the leaves, especially of deciduous *Magnolia* species, when grown under glass. Heavy infestations of magnolia scale (*Neolecanium cornuparyum*) have been reported (Adams, 1972; Gumeringer, 1989). The extent of their natural protection makes it very difficult to eradicate them by spraying. Scales of various types will infest twigs. Overwintering scales can be controlled with dormant oil applied in the spring. Where fumigation is feasible a smoke generator or electrically heated fumigator is recommended (Treseder, 1978).

6.2.4 *Propagation of* Magnolia

6.2.4.1 *Seed sowing and seedling transplantation*

A simple approach to propagation or reproduction of *Magnolia* is from seed. For example, summer-flowering *M. sinensis* and other members of the section *Oyama* are usually prolific seed bearers. In China, seed sowing is a popular method for commercial cultivation of medicinal *M. liliflora*, as Biond Magnolia Flower (Flos Magnoliae).

Many *Magnolia* species are prolific seed producers, and good seed crops are usually produced every year. For example, trees of *M. grandiflora* as young as 10 years can produce seed, but optimum seed production does not occur until age 25. Cleaned seeds range from 12 800 to 15 000/kg. Seed viability averages about 50%. The relatively heavy seeds are disseminated by birds and mammals, but some may be spread by heavy rains (Adams, 1972). *M. grandiflora* is pollinated by insects (Olson *et al.*, 1974; Vines, 1960).

In order to harvest *Magnolia* seed, it is important to know the seed-ripening date of each species. For example, the seeds of *M. officinalis* and *M. hypoleuca* can be collected in late September, as soon as the carpels on the fruit cones begin to split longitudinally to reveal the bright orange, scarlet, or crimson seeds. Delay in harvesting may result in rapid losses to squirrels and other seed-eating rodents and birds. After windrow, the collected seeds can be plunged in cold water and rubbed to take off the arillus of the seeds. The seeds obtained must be indoor seasoned, not sun dried. The seasoned seeds should then be deposited and stored with clean sand in a humid and low-temperature location.

The seeds may be stored in a refrigerator without removing the fleshy coating until February and then sown immediately after cleaning, if it is not possible to clean and treat the seeds soon after harvesting. The cleaned or uncleaned seeds may also be stored immediately in small polythene bags and kept in the lower section of a domestic refrigerator for about 100 days. *Magnolia* seeds tend to lose their viability after six months of storage in normal conditions. Sometimes germination is delayed for over a year from the date of sowing, but rapid germination is usually assured by vernalizing the seed for about 100 days at 2–5 °C (Treseder, 1978).

In practice, the seeds of *M. grandiflora* usually germinate the first or second spring following seedfall; germination is epigeal. The best natural seedbed is a rich, moist soil protected by litter. Even though viable, seeds rarely germinate under the parent tree because of reported inhibitory effects (Blaisdell *et al.*, 1973).

The most economic and effective approach to sowing and cultivation is forced germination in greenhouse and seedling transplantation in the field, rather than direct seeding in the field. The four steps of the approach are as follows: (1) sow the deposited seeds densely into a sandy bed in greenhouse, with a 2 cm thickness of cover sand over the seeds; (2) water the sandy bed daily to keep it wet; (3) transplant each seedling in full-bud stage into a paper cap with soil, and keep the seedling in normal rooting for a month; and (4) transplant seedlings with soil attached into field. Partial shade is beneficial for the first two years of seedling growth. The advantages of this approach over the direct-seeding method are that seed is saved (about 50%) and the emergence of seedlings is accelerated (about one month in advance). For commercial cultivation in China, the approach has proved successful and it is reported that the cultivated trees can reach to 3 m in height and 6 cm in thickness in three years (Liu *et al.*, 1997).

It has been reported that only a small percentage of fruit cones of *Magnolia* develop perfectly in English gardens. This is particularly so with the precocious-flowering species and hybrids, and may be associated with the effect of weather on early insect activity and the susceptibility of *Magnolia* blossoms to frost damage. On the other hand, plants raised from seed normally take longer to attain flowering age than those raised vegetatively as described below, and there is often a considerable variation even between seedlings raised from seed borne within the same carpel (Treseder, 1978).

6.2.4.2 *Cutting*

Cutting is another type of propagation that can be used for many *Magnolia* species, e.g., *M. liliflora*, which fails in attempts to cultivate it commercially with the seed sowing and seedling transplanting approach described above. *Magnolia* does not normally root from dormant hardwood cuttings, but they can be rooted from soft or half-ripe cuttings in the summer. The most suitable periods for cutting propagation for a particular species vary in different locations; for example, the period for *M. liliflora* is May to June in Shanghai and June to July in Beijing, China. The earlier cuttings can be rooted better, since cutting propagated late tend to perish during the first winter after having formed roots.

The rooting medium may be pure sharp sand or a mixture of moss peat and sand. Vermiculite and perlite can be used instead of sand and have the great advantage over sand of being relatively light weight, thus saving time and energy where large-scale propagation is entailed.

The propagation bench should have bottom heat supplied by under-bench hot-water pipes, hot air ducts, or electric soil-heating cables embedded in the drainage layer, to provide a constant temperature of 21–24 °C within the rooting medium. Root initiation can be ascertained by periodic testing of the cuttings. As soon as good clusters of roots have emerged from their bases, they may be lifted and potted. The size of pot selected is only just large enough to accommodate the young roots (Treseder, 1978).

6.2.4.3 *Layering*

Layering is the oldest method of vegetative propagation for woody plants, which do not root spontaneously from branches that have been inserted or driven into the ground. The layering of *Magnolia* has been extensively practiced in nurseries, especially in the Netherlands and Belgium, since the introduction of the Soulangiana hybrids in the 1830s. In the Boskoop nurseries the mother plants or stool bushes (on their own roots) are set out in plots of very heavily manured peaty soil. They are kept meticulously clear of weeds and are mounded in January with cow manure overlaid with leaf-mold.

The long, whippy young shoots that develop are arched outward and downward in August or September, so that portions of them may be buried to induce root formation while the tips are bent upward, supported with bamboo canes. Sometimes it is desirable to layer a *Magnolia* that has no branches close to the ground. In this case, it is necessary to introduce the rooting medium wherever a suitable branch happens to be growing.

6.2.4.4 *Budding*

The budding of *Magnolia* is used extensively in Japan and South Korea. The early successful propagation of *Magnolia* species (*M. kobus* and *M. salicifolia*) by budding was reported by Mundey in 1952.

The seeds of *M. kobus* are drilled about 8 cm apart in rows 15–23 cm apart. Dormant growth buds are cut from half-ripened shoots off the mother trees and the leaf blades are removed before the buds are inserted into vertical "T" cuts on the north sides of the stems of two-year seedlings during warm weather at the end of September. The buds are wrapped with polythene strips and banked up with soil to prevent sun scorch. In the following March the stocks are headed back to a point close to the successful unions before the "buds" begin to swell. Growths of 90–120 cm or more during the first season are not unusual (Treseder, 1978).

Mundey also discussed the considerable time taken to reach flowering maturity, and suggested that a required carbon/nitrogen ratio had to be reached before flower bud initiation could occur.

6.2.4.5 *Grafting*

In practice, grafting has been a routine and effective method of propagation of *Magnolia*. It may be carried out from about mid-August up to the time of growth commencement in May. Better results can be obtained from stocks that have developed hardwood. This contains a thick layer of cambium, those actively dividing cells responsible for callusing and healing.

Grafting should be done with dormant cions. The tree in which the cions are set need not be dormant; in fact, grafts set up to the time of blossoming should grow, provided the cions are dormant. The best time for top working is just before the bark slips.

Cions should be cut from wood of the past season's growth. Better results are from short woody spurs, which are closely ringed with petiolar scars, than from strong young shoots of the current season's growth. With late summer grafting it is usual to reduce the leaves on the cions to half their length to check transpiration.

It has been observed that the two Japanese species *M. kobus* and *M. hypoleuca* provide good understocks of Asian *Magnolia*, while *M. acuminata* is favored in both American and Asian species. *M. sargentiana* also provides a reliable source of understocks. However, only seedlings and rooted cuttings of *M. gradiflora* can be used as understocks for grafting the cultivars of *M. gradiflora* (Treseder, 1978; Hartman *et al.*, 1990).

The understocks must be kept well watered during dry weather and, when the union has been accomplished, all unwanted portions may be cut away gradually. The point of severance from the mother plant must be several inches below the union so that a sizeable snag remains to be pruned away some time after the scion has started into growth, nurtured by the root system of the understock. The final severence is usually best delayed until the end of the following growing season, about the time of normal leaf fall (Treseder, 1978).

Bud grafting or chip-budding is an adaptation of budding that is more akin to grafting. It has the advantage over budding and stem-grafting that it may be practiced

Table 6.6 Comparison of graftings of *Magnolia* (as cions)

Cion	Understock	Grafting time	Grafting method	Survival (%)
M. sprengeri	Michelia champaca	February 1983	Cut	100
M. acuminata var. subcordata	M. denudata	February 1983	Cut	66.7
M. cylindrica	Michelia champaca	February 1983	Cut	100
M. biondii	M. cylindrica	February 1983	Cut	80
M. × soulangeana	Michelia macclurei	February 1985	Cut	85
M. × soulangeana	Michelia champaca	January 1985	Cut	77.5
M. × soulangeana	Michelia platypetala	January 1985	Cut	85.7
M. tripetala	M. cylindrica	February 1985	Cut	50
M. × soulangeana	Michelia champaca	April 1985	Cut	1
M. × soulangeana	Michelia champaca	September 1985	Cut	87.5
M. tripetala	M. officinalis	March 1986	Cut	0
M. tripetala	Manglietia megaphylla	March 1986	Cut	25
M. sieboldii	Michelia champaca	February 1990	Side	50
M. biondii	M. denudata	February 1990	Side	62.5
M. sprengeri	Michelia platypetala	February 1990	Side	58.8
M. tripetala	Michelia champaca	March 1990	Side	10
M. sieboldii	Michelia champaca	February 1991	Bud	36.4
M. zenii	Michelia champaca	February 1991	Bud	50
M. amoena	Michelia champaca	February 1991	Bud	40
M. cylindrica	Michelia champaca	February 1991	Bud	70.8
M. tripetala	Michelia champaca	February 1991	Bud	66.7
M. biondii	Michelia champaca	February 1991	Bud	55
M. grandiflora	Michelia champaca	March 1991	Bud	90

at almost any time of the year, as and when propagating material becomes available. This technique has been developed in America over several decades by fruit- and nut-tree propagators. It is particularly useful for propagating hardwood material, when the bark is tight and when growth conditions are unsuitable for orthodox "T"-budding (Tresedey, 1978).

Experiments with a number of combinations of three grafting methods (cut-grafting, side-grafting, and bud-grafting), different species and varieties of Magnoliaceae as cions and understocks, and different grafting times have been undertaken in South China Botanical Garden, Academia Sinica, Guangzhou, China (Chen and Zeng, 1998). The performances of 23 *Magnolia* species (as cions) and 7 Magnoliaceae species (as understocks) are given in Table 6.6. The results show that both intrageneric and intergeneric grafting of *Magnolia* species can be done if the grafting method and time are suitable. Two fully successful examples (100% survival) are intergeneric, i.e. *Magnolia sprengeri–Michelia champaca* and *Magnolia cylindria–Michelia champaca*, by means of cut-grafting. Other combinations also indicate that intergeneric grafting within the family Magnoliaceae is efficient. The grafting time, however, is a limiting factor. For example, the grafting survivals of *Magnolia tripetala* in February 1985 (50%) and February 1991 (66.7%) were higher than those in March 1986 (0%) and March 1986 (25%), no matter whether the understock was *Magnolia* or not. Interestingly, the grafting of *Magnolia* × *soulangeana* with *Michelia champaca* was favorable when done in late September 1985 (87.5% survival).

6.2.5 *Microculture of* Magnolia

The potential utilization of microculture in the propagation and study of *Magnolia* is considerable, but this potential has yet to be realized. The *in vitro* performance of *Magnolia* is often poor relative to that of most other microcultured plants, largely owing to the absence of optimized *in vitro* methodologies.

Protocols for the rapid and reliable cloning of *Magnolia* genotypes are critical to efficient micropropagation and would be an important component of a biological or phytochemical study. These techniques would provide the consistent tissues necessary for replicated analyses. Genotypes of interest can be established in microculture as sterile shoot cultures from which uniform organs or shoots can be obtained independently of seasonal variation.

In general, *Magnolia* species possess high levels of phenolic compounds (Biedermann, 1987). Phenolic exudates are a recurring problem in the microculture of many woody plants (McCown, 2000), and likely reduce the efficiency of *Magnolia* microculture. *Magnolia* species also possess relatively large *in vitro* shoots and leaves, a fact that greatly limits the number of shoots that can be grown per microculture vessel.

The rooting and establishment of *Magnolia* microcuttings appears to be a greater problem than is *in vitro* multiplication. The rooting of *Magnolia* cuttings is often difficult (Dirr and Heuser, 1987), but some studies have shown encouraging results. Maene and Debergh (1985) found that supplementing established cultures with liquid media prior to *ex vitro* rooting significantly increased rooting percentage. Kamenicka (1998) analyzed the effects of different carbohydrate sources in the media and found that fructose, mannose, and xylose promoted the best *in vitro* rooting over a period of 13 weeks.

Most reports of *Magnolia* microculture involve the establishment and multiplication of genotypes on conventional media/hormone formulations utilized for a range of plant genera. The definition of media/hormone responses, particularly cytokinin response curves, is critical to the development of an effective microculture protocol. Biedermann (1987) cultured several genotypes of *Magnolia* on a number of media formulations and found that *Magnolia* performed better on media containing relatively low salt concentrations. This suggests that the widely used Murashige and Skoog (MS) (Murashige and Skoog, 1962) medium may be inferior to lower-salt media such as Anderson's Medium (AR) (Anderson, 1984) and Woody Plant Medium (WPM) (Lloyd and McCown, 1981) for *Magnolia* microculture. In China, the first reported medicinal *Magnolia in vitro* was *M. officinalis*, for which a modified MS medium was used (Chen, 1986; Zhong *et al.*, 1984).

The aforementioned concerns indicate that shoot culture may not be an effective means of generating large numbers of a *Magnolia* genotype. However, substantial numbers may be possible through somatic embryogenesis. Work with *M. virginiana*, *M. fraseri*, and *M. acuminata* (Merkle and Wiecko, 1990) indicate that somatic embryogenesis could potentially perform this role if *Magnolia* genotypes are needed in great quantity.

Another potential application of microculture in *Magnolia* propagation may be to maintain stock plant juvenility. As an alternative or supplement to micropropagation, microculture may serve as a means of supplying highly juvenile stock plants from which quality softwood cuttings can be obtained. The juvenility obtained from *in vitro* culture can improve stock plant vigor and the rooting percentage of cuttings (George, 1993).

The problems related to *Magnolia* microculture should not be a deterrent to developing efficient *in vitro* systems for this genus. Economically feasible systems have been developed for other woody plants that were initially recalcitrant *in vitro*, including *Kalmia* (Lloyd and McCown, 1981) *Rhododendron* (McCown and Lloyd, 1983) and *Syringa* (Pierik *et al.*, 1988). These three genera are now in widespread commercial microculture. Other woody genera for which optimized *in vitro* methodology has been developed and which are now commonly produced through micropropagation include *Amelanchier*, *Betula*, *Populus*, and *Ulmus* (McCown, 2000).

The relatively modest success achieved to date in the microculture of *Magnolia* holds significant promise for the future application of this technology to the propagation and study of this genus. Thorough investigation to determine the optimum protocol for *in vitro* establishment, multiplication, and rooting will be required before the widespread microculture of *Magnolia* becomes a reality.

Acknowledgments

We thank Dr David Boufford for helpful comments, and Ms Hong Jin, Ms Yunping Huang, Ms Xiaoyan Zhang and Mr Yunhai Pu for providing references or technical assistance. This work was partially supported by a grant from the National Natural Science Foundation of China (NSFC) for Distinguished Youth Scholars (39825104) to S. Shi.

References

Adams, D.L. (1972) *Natural regeneration following four treatments of slash on clear-cut areas of lodgepole pine—a case history*. Stn. Not. No. 19. Moscow, ID: University of Idaho, Forestry, Wildlife, and Range Experiment Station.

Anderson, W.C. (1984) A revised tissue culture medium for shoot multiplication of *Rhododendron*. *J. Am. Soc. Hort. Sci.*, 109, 343–347.

Azuma, H., Thien, L.B. and Kawano, S. (1999) Molecular phylogeny of *Magnolia* inferred from cpDNA sequences and evolutionary divergence of the floral scents. *J. Plant Res.*, 112, 291–306.

Bande, M.B. and Prakash, S.U. (1986) The Tertiary flora of Southeast Asia with remarks on its paleoenvironment and phytogeography of the Indo-Malayan region. *Rev. Paleobot. Palynol.*, 49, 203–233.

Bell, W.A. (1957) Flora of the Upper Cretaceous Nanaimo Group Vancouver Island, British Columbia. *Geol. Surv. Can. Mem.*, 285, 1–331.

Bennett, F.A. (1965) Southern magnolia (*Magnolia grandiflora* L.). In *Silvics of Forest Trees of the United States*, edited by H.A. Fowells, pp. 27–276. Agriculture Handbook 271. Washington, DC: US Department of Agriculture.

Biedermann, I.E. (1987) Factors affecting establishment and development of *Magnolia* hybrids *in vitro*. *Acta Hort.*, 212, 625–629.

Blaisdell, R.S., Wooten, J. and Godfrey, R.K. (1973) The role of magnolia and beech in forest processes in the Tallahassee, Florida, Thomasville, Georgia area. *Proceedings of Thirteenth Annual Tall Timbers Fire Ecology Conference*, pp. 363–397.

Boufford, D.E. (1992) Affinities in the floras of Taiwan and eastern North America. In *Phytogeography and Botanical Inventory of Taiwan*, edited by Ching-I Peng, pp. 1–16. Academia Sinica Monograph Series No. 12. Taipei, Taiwan: Institute of Botany.

Boufford, D.E. and Spongberg, S.A. (1983) Eastern Asian-eastern North American phytogeographical relationships—a history from the time of Linnaeus to the twentieth century. *Ann. Mo. Bot. Gard.*, 70, 423–439.

Callaway, D.J. (1994) *The World of Magnolias*. Portland, OR: Timber Press.

Chen, B.L. and Nooteboom, H.P. (1993) Notes on Magnoliaceae III: the Magnoliaceae of China. *Ann. Mo. Bot. Gard.*, 80, 999–1104.

Chen, W. and Zeng, Q. (1998) The grafting propagation of Magnoliaceae. *J. Trop. Subtrop. Bot.*, 6, 70–78.

Chen, Z. (1986) *Tissue Culture of Trees*. Beijing: Higher Education Publishing House.

Dandy, J.E. (1927) The genera of Magnolieae. *Kew Bull.*, 1927, 257–264.

Dandy, J.E. (1964) Magnoliaceae. In *The Genera of Flowering Plants, Angiospermae 1*, edited by J. Hutchinson, pp. 54–56. Oxford: Clarendon Press.

Dandy, J.E. (1974) Magnoliaceae taxonomy. In *Magnoliaceae Juss. World Pollen Spore Flora*, vol. 3, edited by J. Praglowski. London: Almqvist & Faber.

Dandy, J.E. (1978) A revised survey of the genus *Magnolia* together with *Manglietia* and *Michelia*. In *Magnolias*, edited by N.G. Treseder, pp. 29–37. London and Boston: Faber and Faber.

Diels, L. (1900) Die flora von Central-China. *Bot. Jahrb. Syst. Pflanzengesch. Pflanzengeogr.*, 29, 167–659.

Dilcher, D.L. and Crane, P.R. (1984) *Archacanthus*: an early angiosperm from the Cenomanian of the western interior of North America. *Ann. Mo. Bot. Gard.*, 71, 351–383.

Dirr, M.A. and Heuser, C.W. (1987) *The Reference Manual of Woody Plant Propagation*. Athens, GA: Varsity Press.

Doyle, J.A. and Hickey, L.J. (1976) Pollen and leaves from the mid-Cretaceous Potomac Group and their bearing on early angiosperm evolution. In *Origin and Early Evolution of Angiosperms*, edited by C.B. Beck, pp. 139–206. New York: Columbia University Press.

Duncan, W.H. and Duncan, M.B. (1988) *Trees of the Southeastern United States*. Athens, GA: The University of Georgia Press.

Elias, T.S. (1980) *The Complete Trees of North America*. New York: Times Mirror Magazines.

Engler, A. (1879) *Versuch einer Entwicklungsgeschichte der Pflanzenwelt, inbesondere der Florengebiete seit der Tertiärperiode, 1. Die extratropischen Gebiete der nordlichen Hemisphäre*. Leipzig: Wilhelm Engelmann.

Fernald, M.L. (1931) Specific segregations and identities in some floras of eastern North America and the Old World. *Rhodora*, 33, 25–63.

Figlar, R.B. (2000) Proleptic branches initiation in *Michelia* and *Magnolia* subgenus *Yulania* provides basis for combinations in Subfamily Magnolioideae. In *Proceedings of the International Symposium on the Family of Magnoliaceae*, edited by Y.H. Liu, pp. 14–25. Beijing: Science Press.

Frodin, D.G. and Govaerts, R. (1996) *World Checklist and Bibliography of Magnoliaceae*. Kew, UK: Royal Botanic Gardens.

George, E.F. (1993) *Plant Propagation by Tissue Culture, Part I. The Technology*, pp. 239–241. Edington: Exegetics Limited.

Glitzenstein, J.S., Harcombe, P.A. and Streng, D.R. (1986) Disturbance, succession, and maintenance of species diversity in an east Texas forest. *Ecol. Monogr.* 56, 243–258.

Godfrey, R.K. (1988) *Trees, shrubs, and woody vines of northern Florida and adjacent Georgia and Alabama*. Athens, GA: The University of Georgia Press.

Gray, A. (1840) Dr. Siebold, Flora Japonica; sectio prima. Plantas ornatui vel usui inservientes; digessit Dr. J.G. Zuccarini: fasc. 1–10, fol. *Am. J. Sci.*, 39, 175–176.

GRIN database (2001) available online at http://www.ars-grin.gov/npgs/tax/index.html

Gumeringer, K. (1989) *Magnolia (Magnoliaceae* and *Annonaceae). Forest World*, 5, 4–45.

Guo, C.L. and Huang, L.L. (1992) A new species of medicinal *Magnolia* in Hubei, China: *M. elliptigemmata. J. Wuhan Bot. Res.*, 10, 325–327.

Guo, X. (1986) General features and successions of Cretaceous floras in China and North Hemisphere. *Acta Palaeobot. Palynol. Sinica*, 1, 31–45.

Guo, X. and Sun, B. (1989) Neogene megafossil and climatic events in China. In *Proceedings of International Symposium on Pacific Neogene Continental and Marine Events*, edited by R. Tsuchi and Q. Lin, pp. 91–102. Nanjing: Nanjing University Press.

Halenius, J. (1750) Plantae Rariores Camschatcenses. *Thesis.* University of Uppsala, Uppsala, Sweden.

Hara, H. (1966) Magnoliaceae. In *Flora of the Eastern Himalaya*, edited by H. Hara, pp. 95–96. Tokyo: The University of Tokyo Press.

Hartman, H.T., Kester, D.E. and Davies, F. (1990) *Plant Propagation: Principles and Practices.* Englewood Cliffs, NJ: Prentice-Hall.

Hsu, J., Jang, D. and Yang, H. (1974) Sporo-pollen assemblage and geological age of the Lower Xinminbu Formation of Chiuchuan, Kansu. *Acta Bot. Sinica*, 16, 365–379.

Hu, H.H. (1935) A comparison of the ligneous flora of China and eastern North America. *Bull. Chinese Bot. Soc.*, 1, 79–97.

Hu, H.H. and Chaney, R.W. (1940) A Miocene flora from Shantung Province, China. *Palaeontogr. Sinica, NSA*, 1, 1–147.

Jiang, D., He, Z. and Dong, K. (1988) The Early Cretaceous sporo-pollen assemblage from the Tarim Pendi, Xinjiang. *Acta Bot. Sinica*, 30, 430–440.

Kamenicka, A. (1998) Influence of selected carbohydrates on rhizogenesis of shoots saucer magnolia *in vitro. Acta Physiolo. Plant.*, 20, 425–429.

Keng, H. (1978) The delimitation of the genus *Magnolia* (Magnoliaceae). *Garden Bulletin Straits Settlement*, 31, 127–131.

Lape, F. (1966) Note. *Newsl. Am. Magnolia Soc.*, 3, 1–6.

Law, Y.W. (1984) A preliminary study on the taxonomy of the family Magnoliaceae. *Acta Phytotaxon. Sinica*, 22, 89–109.

Law, Y.W. (1990) Yulan. In: *Chinese Flowers*, edited by Chen, J., pp. 186–191. Beijing: China Architecture Press.

Law, Y.W. (1996) Magnoliaceae. In *Flora Reipublicae Popularis Sinicae*, vol. 30(1), edited by Agendae Academiae Sinicae, pp. 87–282. Beijing: Science Press.

Law, Y.W. (1997) W*oonyoungia*, a new genus of Magnoliaceae from China. *Bull. Bot. Res.*, 17, 353–356.

Law, Y.W. (1999) Origin, evolution and geographical distribution of Magnoliaceae. In *The Geography of Spermatophyta Families and Genera*, edited by A.M. Lu, pp. 65–74. Beijing: Science Press.

Lay, D.W. (1957) Browse quality and the effects of prescribed burning in southern pine forests. *J. Forestry*, 55, 342–347.

Leach, D.L. (1973) Note. *Newsl. Am. Magnolia Soc.*, 9, 3–4.

Li, H.L. (1952) Floristic relationships between eastern Asia and eastern North America. *Transactions of the American Philosophical Society Held at Philadelphia for Promoting Useful Knowledge*, 42, 371–429.

Li, M., Song, Z. and Li, Z. (1978) A few Cretaceous-Tertiary sporo-pollen assemblages from Jiang-Han Plain. In *Acta Nanjing Inst. Geol. Palaeontol.*, 31, Beijing: Science Press.

Li, X.X. (1995) *Fossil Floras of China Through the Geological Ages*, p. 432. Guangzhou: Guangdong Science and Technology Press.

Little, S., Moorhead, G.R and Somes, H.A. (1958) *Forestry and deer in the Pine Region of New Jersey*. Station Paper No. 109. Upper Darby, PA: US Department of Agriculture, Forest Service, Northeastern Forest Experiment Station.

Liu, Y.H. (2000) Studies on the phylogeny of Magnoliaceae. In *Proceedings of the International Symposium on the Family of Magnoliaceae*, edited by Y.H. Liu, pp. 3–13. Beijing: Science Press.

Liu, Y.H., Zhou, R. and Zeng, Q. (1997) Study on *ex situ* conservation of Magnoliaceae and its rare and endangered species. *J. Trop. Subtrop. Bot.*, 5, 120–127.

Lloyd, G.B. and McCown, B.H. (1981) Commercially-feasible micropropagation of mountain laurel, *Kalmia latifolia* by use of shoot tip culture. *Int. Plant Propagator's Soc. Proc.*, 30, 421–427.

Maene, L. and Debergh, P. (1985) Liquid medium additions to established tissue cultures to improve elongation and rooting *in vivo. Plant Cell, Tissue, Organ Culture*, 5, 23–33.

Maisenhelder, L.C. (1970) *Magnolia*. In *American Woods*, FS-245, p. 7. Washington, DC: USDA Forest Service.

McCown, B.H. (2000) Recalcitrance of woody and herbaceous perennial plants: dealing with genetic predeterminism. *In Vitro Cell Dev. Biol. Plant*, 36, 149–154.

McCown, B.H. and Lloyd, G.B. (1983) A survey of the response of *Rhododendron* to *in vitro* culture. *Plant Cell, Tissue, Organ Culture*, 2, 77–85.

Merkle, S.A. and Wiecko, A.T. (1990) Somatic embryogenesis in three *Magnolia* species. *J. Am. Soc. Hort. Sci.*, 115, 858–860.

Miquel, F.A.W. (1867) Sur les affinites de la flore du Japon. *Adansonia*, 8, 132–153.

Murashige, T. and Skoog, F.S. (1962) A revised medium for rapid growth and bioassays with tobacco tissue cultures. *Physiol. Plant.*, 15, 473–497.

Myers, R. and White, D.L. (1987) Landscape history and changes in sandhill vegetation in north-central and south-central Florida. *Bull. Torrey Bot. Club*, 114, 21–32.

Nooteboom, H.P. (1985) Notes on Magnoliaceae I. *Blumea*, 31, 66–87.

Nooteboom, H.P. (2000) Different looks at the classification of the Magnoliaceae. In *Proceedings of the International Symposium on the Family of Magnoliaceae*, edited by Y.H. Liu, pp. 26–37. Beijing: Science Press.

Nuttall, T. (1818) *The Genera of North American Plants and a Catalogue of the Species to the Year 1817*. Printed for the author by D. Heartt, Philadelphia.

Olson, D.F., Barnes, R.L. and Jones, L. (1974) *Magnolia* L. In *Seeds of Woody Plants in the United States*, edited by C.S. Schopmeyer, pp. 52–530. Agriculture Handbook No. 450. Washington DC: US Department of Agriculture, Forest Service.

Page, V.M. (1970) Angiosperm wood from the upper Cretaceous of California III. *Am. J. Bot.*, 57, 1139–1144.

Pierik, R.L., Steegmans, H.H., Elias, A.A., Stiekema, O.T. and Van Der Velde, A.J. (1988) Vegetative propagation of *Syringa vulgaris* L. *in vitro*. *Acta Hort.*, 226, 195–204.

Priester, D.S. (1990) *Magnolia virginiana* L. sweetbay. In *Silvics of North America*, vol. 2, *Hardwoods*, edited by R.M. Burns and B.H. Honkala, pp. 445–448. Washington DC: US Department of Agriculture, Forest Service.

Pursh, F. (1814) *Flora Americae Septentrionalis; or a Systematic Arrangement and Description of the Plants of North America*. 2 vols. London: White, Cochrane.

Qiu, Y.L., Parks, C.R. and Chase, M.W. (1995) Molecular divergence in the eastern Asia-eastern North America disjunct section *Rytidospermum* of *Magnolia* (Magnoliaceae). *Am. J. Bot.*, 82, 1589–1598.

Raven, P.H. (1972) Plant species disjunctions: a summary. *Ann. Mo. Bot. Gard.*, 59, 234–246.

Sewell, M.M., Qiu, Y.L., Parks, C.R. and Chase, M.W. (1993) Genetic evidence for trace paternal transmission of plastids in *Liriodendron* and *Magnolia* (Magnoliaceae). *Am. J. Bot.*, 80, 854–858.

Shi, S., Jin, H., Zhong, Y., He, X., Huang, Y., Tan, F. and Boufford, D.E. (2000) Phylogenetic relationships of the Magnoliaceae inferred from cpDNA *mat*K sequences. *Theor. Appl. Genet.*, 101, 925–930.

Simpson, B.J. (1988) *A Field Guide to Texas Trees*. Austin, TX: Texas Monthly Press.

Song, Z. (1986) A review on the study of Early Cretaceous angiosperm pollen in China. *Acta Micropalaeontol. Sinica*, 3, 373–386.

Song, Z., Zheng, Y., Liu, J., Ye, P., Wang, C. and Zhou, S. (1981) *Cretaceous–Tertiary Palynological Assemblages from Jiangsu*, pp. 1–269. Beijing: Geographical Publishing House.

Song, Z., Wang, X. and Zhong, L. (1992) Early Cretaceous angiospermous pollen from Baihedong Formation of Sanshui Basin, Guangdong. *Acta Palaeontol. Sinica*, 31, 501–512.

Staff of the L.H. Bailey Hortorium, Cornell University. (1976) *Hortus Third*. New York: Macmillan.

Sun, X. (1979) Palynofloristical investigation on the Late Cretaceous and Paleocene of China. *Acta Phytotaxon. Sinica*, 17, 8–23.

Takhtajan, A. (1974) Magnoliophyta Fossilia. *URSS*, 1, 5–21.

Tao, J. and Du, N. (1982) Neogene flora of Tengchong Basin in western Yunnan, China. *Acta Bot. Sinica*, 29, 649–655.

Tao, J. and Kong, Z. (1973) The fossil florula and sporo-pollen assemblage of Shang-in coal series of Erhyuan, Yunnan. *Acta Bot. Sinica*, 15, 120–126.

Tao, J. and Zhang, C. (1992) Two angiosperm reproductive organs from the Early Cretaceous of China. *Acta Phytotaxon. Sinica*, 30, 423–426.

Taylor, D.W. (1990) Paleobiogeographic relationships of angiosperms from the Cretaceous and Early Tertiary of the North American area. *Bot. Rev.*, 56, 279–417.

Thorne, R.F. (1972) Major disjunctions in the geographical ranges of seed plants. *Quart. Rev. Biol*, 47, 365–411.

Thunberg, C.P. (1784) *Flora Japonica*. Leipzig: I.G. Mulleriano.

Tiffney, B.H. (1977) Fruits and seeds of the Brandon Lignite: Magnoliaceae. *Bot. J. Linn. Soc.*, 75, 299–322.

Treseder, N.G. (1978) *Magnolias*. London and Boston: Faber & Faber.

Tripp, K.E. and Raulston, J.C. (1995) *The Year in Trees: Superb Woody Plants for Four-season Gardens*. Portland, OR: Timber Press.

Ueda, K., Yamashita, J. and Tamura, M.N. (2000) Molecular phylogeny of the Magnoliaceae. In *Proceedings of the International Symposium on the Family of Magnoliaceae*, edited by Y.H. Liu, pp. 205–209. Beijing: Science Press.

Vines, R.A. (1960) *Trees, Shrubs, and Woody Vines of the Southwest*. Austin, TX: University of Texas Press.

Wang, M.C. and Min, C.L. (1992) A new species of *Magnolia* from Shanxi. *Acta Botanica Boreali-Occidentali Sinicae*, 12: 85.

Wiersema, J.H. and Leon, B. (1999) *World Economic Plants: A Standard Reference*. Boca Raton, FL: CRC Press.

Wood, C.E., Jr. (1971) Some floristic relationships between the Southern Appalachians and western North America. In *The Distributional History of the Biota of the Southern Appalachians, Part II. Flora*, edited by P.C. Holt, pp. 331–404. Virginia Research Division Monograph 2. Blacksburg, VA: Virginia Polytechnic Institute and State University.

Ye, D. and Zhang, X. (1990) *Cretaceous in Oil and Gas Bearing Areas of Northern China*, pp. 1–135. Beijing: Petroleum Industry Press.

Zhao, Z.Z. and Xie, J. (1987) A new species and a new nomenclature of the medicinal Xin-Yi (*Flos magnoliae*). *Acta Pharmacol. Sinica*, 22, 777–780.

Zhong, W., Liu, S. and Yang, G. (1984). A preliminary study *in vitro* of *M. officinalis*. *Chinese Medicinal Herbs*, 15, 31–33.

Index

References to figures occurring outside their relevant page ranges are in *italic*; references to tables are in **bold**.

Printed and bound by CPI Group (UK) Ltd, Croydon, CR0 4YY

23/10/2024

01778248-0009